Mosses of the
Gulf South

William Dean Reese

Mosses of the Gulf South

**FROM THE
RIO GRANDE
TO THE
APALACHICOLA**

**Louisiana
State
University
Press**
*Baton Rouge and
London*

Copyright © 1984 by Louisiana State University Press
All rights reserved
Manufactured in the United States of America

Designer: Barbara Werden
Typeface: Linotron Optima
Typesetter: G & S Typesetters, Inc.
Printer: Thomson-Shore, Inc.
Binder: John H. Dekker & Sons

Library of Congress Cataloging in Publication Data

Reese, William Dean.
  Mosses of the Gulf South.

  Bibliography: p.
  Includes index.
  1. Mosses—Gulf States—Identification.   I. Title.
QK541.5.G84R43   1983   588'.10976   83-889
ISBN 0-8071-1110-4

To Ruth Schornherst Breen

Mofs is a Plant, that the wifeft of
Kings thought neither unworthy
his fpeculation, nor his pen,
and though amongft Plants it be
in bulk one of the fmalleft, yet it
is not the leaft confiderable: For,
as to its fhape, it may compare
for the beauty of it with any
Plant that grows.

Robert Hooke
Micrographia, 1665

Contents

Illustrations

Figures

Tables

Preface

Mosses are small green plants, many of great delicacy and beauty, which play a modest role in nature's economy. Unnoticed by many visitors to their habitats because of their unpretentious way of life, mosses are nonetheless almost ubiquitous in land and freshwater areas of the world. The humid forests of the Gulf Coast are rich in mosses and even the cypress swamps of Louisiana and the grassy prairies of Texas support varied populations of mosses.

Relatively few people have studied mosses along the Gulf Coast, and there is still much to be learned about the mosses that grow there. This book, which is both an introduction to the science of muscology and a manual of the mosses of the Gulf Coast, was written with two main purposes in mind: to provide a convenient means for identification of Gulf Coast mosses and to stimulate interest in the study of mosses.

The area covered by the book extends from the Rio Grande River in the west to the Apalachicola River in the east. Coverage by states includes the coastal-plain area of Texas, all of Louisiana, the southern portions of Mississippi and Alabama, and western Florida. The book will also be useful in adjoining regions.

Many people share responsibility with me in one way or another for developing and writing this book. Foremost is Ruth Schornherst Breen, whose patience, enthusiasm, and cheerfully optimistic approach to scholarship and teaching encouraged and guided my interest in mosses. I was her student for four years at Florida State University. I also owe much to Robert K. Godfrey for his friendly counsel and companionship and for providing invaluable opportunities for field work during my student years in Florida. Andrew Westling, Raymond Jones, Betty Lemmon, Wayne Guerke, Monte Manuel, Alan Neumann, Charles Allen and Karl Vincent, all former students of mine, and also Tim Morton, Alan Broussard, John Thieret, and Jake Valentine, spent numerous pleasant and rewarding days on field trips with me.

A large portion of the information in this book was gained on these trips. I learned much from each of these people; I thank them all sincerely and hope that they enjoyed the experience and learned as much as I did. Ronald Pursell traveled and collected with me for many weeks under the ardent Gulf Coast sun; he also reviewed the *Fissidens* portion of the book manuscript for me, and I am grateful to him on both counts. Lewis Anderson, Howard Crum, Jerry Snider, and Ann Rushing provided specimens of particular interest for this study and in some cases helped me significantly in other ways too. The University of Southwestern Louisiana has provided me, over the years, with the atmosphere and facilities essential to my work, and my colleagues in the Department of Biology have been understanding and supportive throughout. In the end, sole responsibility for the content and philosophy of this book is mine, and I will welcome having my attention called to errors of fact or omission.

Mosses of the
Gulf South

Part One

INTRODUCTION
TO BRYOLOGY

THE MEANINGS OF *MOSS*

The word *moss* may have different meanings to different people. To Southerners *moss* usually means Spanish moss, known botanically by the name *Tillandsia usneoides*. This graceful epiphyte is a member of the flowering plant family Bromeliaceae (bromeliads) and thus related to the pineapple. Botanically, it is not a moss at all. Reindeer moss, a staple food of the reindeer and caribou in arctic regions, is really a lichen rather than a moss. To a fisherman *moss* may mean the tangled mats of submerged plants that foul the hooks on his lures. To many people the word *moss* merely means any small, low-growing mossy assemblage of plants, including perhaps algae, lichens, and liverworts, as well as a variety of small, low, or creeping plants of other kinds.

For the purposes of this book, the word *moss* is used in its traditional, restricted botanical sense to refer specifically to the small, leafy plants classified together in the formal botanical category of Musci. Similar to mosses in general appearance and habitat, the liverworts and hornworts are traditionally grouped together with mosses under the convenient name *bryophytes*. This book deals only with the mosses.

WHY STUDY MOSSES?

Bryology—the science or study of bryophytes—is not for everyone. In 1982 the total North American membership (U.S.A., Canada, Mexico) of the American Bryological and Lichenological Society, Incorporated, was 325 individuals, and this number included not only muscologists (those who study mosses) but also hepaticologists (those who study liverworts and hornworts) and lichenologists (those who study lichens). The term *bryology* refers to the study of bryophytes in general—mosses, liverworts, and hornworts.

It is not easy to answer the question, "Why study mosses?" Answers from individual bryologists, both amateur and professional, would probably range from the inane ("because they're there") to the highly accurate ("because they make excellent subjects for many types of fundamental research"). Whatever the truth might be in particular cases, many people who study and work with mosses do so because they enjoy working with plants of such delicate beauty and diverse features. Robert Hooke, who is quoted at the beginning of this book, was the first person to comment in writing on the charm of these plants as seen through the microscope. Although we live in an age of high technology, there is still much to be learned from basic studies of plants, including mosses. Not only are mosses rewarding subjects for laboratory investigations, but their study in the field can still also be of value. Even though mosses have been studied more or less formally in North America for more than a hundred years, there is yet a great deal to be learned about the distribution and ecology of North American mosses. Such studies are based primarily upon the collection and identification of mosses and may involve also making observations and measurements in the field as well as in the laboratory. Some parts of North America are very well known bryologically, others only poorly known. Yet there are opportunities for rewarding study of mosses everywhere.

IMPORTANCE OF MOSSES

There are several ways in which mosses are important: in commerce, in the natural economy of nature, in their aesthetic value, and as subjects for scientific study.

As objects of commerce, mosses are not of high significance. Peat moss (*Sphagnum*) is harvested, packaged, and sold, mostly for use as a soil conditioner. Many other kinds of mosses are harvested, often in mixture, from natural habitats and sold primarily for decorative purposes. Such mosses, often dyed shades of green, are commonly seen in planters in such places as restaurants and dentists' offices. There is also a small trade in living mosses for use in terrariums and as a ground cover for potted plants, usually in bonsai arrangements. Thieret (1965) has reviewed the economic uses of mosses in a very interesting and entertaining paper.

Mosses, like all organisms, play complex roles in nature, and it is not a simple matter to state all of the ways in which they are of importance ecologically. Few organisms feed directly upon mosses, although some do, including some insects, certain rodents, and some birds. In general, mosses are of small importance as a direct food source for animals. Some birds, the Carolina wren for example, routinely use mosses in nest construction, as do some rodents. Some mosses are probably of considerable importance in reducing erosion of soil from otherwise bare soil surfaces. Also, mosses play a role in colonizing bare soil and rock surfaces, forming a mat that may ultimately support larger forms of plants. Aquatic mosses, such as *Fontinalis*, may support entire communities of other organisms as well as offering shelter to small aquatic animals.

The Sabine map turtle (*Graptemys ouachitaensis* subsp. *sabinensis*) incorporates mosses in its diet. This turtle lives in streams in eastern Texas and western Louisiana and grazes mosses from logs and rocks in the water. *Fissidens fontanus* and *Amblystegium riparium* have been recovered in some quantity from stomachs of these turtles. It is not known if the turtles eat the moss directly for food or only incidentally in a quest for small animals contained in the mossy vegetation.

The aesthetic value of mosses is easily recognized by anyone who looks for them in their natural habitats. Forming soft, green carpets on tree bases, rocks, and rotting logs, mosses lend a gentling and pleasing touch to the forest scene. The natural beauty of mosses has long been recognized by poets, as, for example, Samuel Taylor Coleridge, who included the following stanza in his "Rime of the Ancient Mariner":

> He kneels at morn, and noon, and eve—
> He hath a cushion plump:
> It is the moss that wholly hides
> The rotted old oak-stump.

Another poet, Richard Henry Dana, dedicated an entire poem to moss, in which he has the moss speak in the first person to the poet! ("The Moss Supplicateth for the Poet"). Many other references to mosses occur in literature, attesting to the enduring recognition of the natural grace and charm of these modest plants.

Mosses are very old plants in the sense that they have existed, probably in much the same form that they have today, longer than virtually any other living land plants. Thus they have great scientific value as a living indication of what early land plants might have been like. Because the majority of mosses probably evolved into their present forms well before separation of the continents (continental drift), the present-day distribution of mosses around the world gives valuable clues as to land-mass movements and ancient vegetation patterns. The ease with which mosses can be cultured in the laboratory, coupled with their unique life cycle, makes them valuable for basic studies in experimental morphology. Among land plants, only liverworts and hornworts share with the mosses the feature of a dominant gametophyte generation. In all other land plants the sporophyte generation is the dominant or conspicuous generation.

HOW TO TELL MOSSES FROM OTHER PLANTS

It is usually easy to recognize that a given plant is a moss rather than a liverwort or hornwort, a lichen or fungus, an alga, or even a vascular plant. In some cases it is not so easy and even experts can be fooled in the field and have to resort to the microscope for certain identification. The following paragraphs will aid in distinguishing mosses from other small plants.

How To Tell Mosses from Fungi

Virtually all mosses have stems bearing small leaves and are green; no fungi have leaves (or even true stems) and few are green. The most common forms of fungi, such as mushrooms, puffballs, bracket fungi, and coral fungi are unlikely to be confused with mosses.

How To Tell Mosses from Lichens

No lichens have leafy stems, although some produce small flakes, or squamules, that may resemble leaves at first glance. Quite a few lichens are green, although usually of a different hue from the green of mosses.

How To Tell Mosses from Algae

Some algae that live in the same habitats as mosses can be taken for mosses at first glance. However, no algae have leafy stems. Instead, algae exist for the most part as individual cells or clusters of cells, or as filaments of cells attached end to end. Most algae are aquatic, but a few are terrestrial; some grow on the bark of trees. Most mosses are terrestrial, but a few are aquatic. *Chara* and *Nitella* are two fairly large aquatic algae that have stems and branches. Neither has leaves and thus should not be mistaken for a moss. The alga *Trentepohlia* often forms conspicuous orange to greenish colonies on tree bases and rocks along the Gulf Coast.

How To Tell Mosses from Vascular Plants

Plants that have conducting tissue—xylem and phloem—in their bodies are called vascular plants. All of the higher plants, including ferns and fern allies, gymnosperms such as pines and cedars, and flowering plants such as dogwood and daylily, are vascular plants. No moss has true xylem and phloem, although some have similar tissues for conduction. The easiest way to distinguish between mosses and small vascular plants is by examining the leaves of the plant in question. Leaves of mosses are (with rare exceptions) only one cell layer thick, other than the midrib; the leaves of vascular plants (again with rare exceptions) are from two to many layers of cells thick. Still, even experts can be fooled. Baron Ferdinand Jacob Heinrich von Mueller, eminent student of Australian flora, described an Australian plant as a new species under the name *Trianthema humillima*, placing it in the flowering plant family *Aizoaceae*. Reexamination of the specimens, however, showed that they really represented a moss, *Gigaspermum repens* (Black 1943–57).

The vascular plant most frequently mistaken for a moss is the fern ally *Selaginella*, often called spikemoss. Some species grow in woodland habitats among true mosses and are delicate and very mosslike in appearance. Their relatively thick, rigid leaves and possession of true roots aid in distinguishing them from mosses. Another fern ally, *Lycopodium*, the club moss, is not likely to be taken for a moss because of its robust stature.

How To Tell Mosses from Liverworts and Hornworts

Mosses, liverworts, and hornworts share many features in common and were long treated by botanists as though they were closely related. While it is still convenient to refer to all of these plants as bryophytes, the three groups are not in fact at all closely related, no more so perhaps than flowering plants are related to ferns. With a little practice it is not usually difficult to tell mosses from liverworts and hornworts, even with the unaided eye.

Virtually all mosses have leafy stems. In contrast, hornworts and many liverworts have thallose bodies, meaning that the plant body is not differentiated into stems and leaves. Instead, the body of hornworts and many liverworts consists of a flattened, leafless, sometimes ribbonlike structure, usually prostrate on the substrate and variously branched. Such plants are easily distinguished from mosses. Most of the other liverworts, however, are leafy, that is, their vegetative body consists of a stem bearing leaves in a manner similar to that of mosses. Even so, mosses differ from leafy liverworts in a number of ways. Table 1 gives comparisons that can be used to distinguish between mosses and leafy liverworts. The list is not exhaustive.

Table 1. Comparison of Mosses and Leafy Liverworts

Mosses	Leafy Liverworts
Plants growing erect or prostrate	Plants almost always growing prostrate
Leaves often with a midrib	Leaves never with a midrib
Leaves not lobed	Leaves often lobed
Leaf cells isodiametric or elongate	Leaf cells isodiametric
Leaves rarely inserted in three rows on stem; mostly spirally inserted	Leaves commonly inserted in three rows on stem—two lateral rows and a third ventral row of usually smaller leaves; never spirally inserted
Leaf cells without oil bodies	Leaf cells commonly with oil bodies
Leaves mostly longer than wide, pointed	Leaves commonly about as long as wide, rarely pointed
Sporophytes mostly tough, persistent	Sporophytes delicate, not long lasting
Capsules usually opening by an operculum; peristome usually present; elaters not produced	Capsules lacking an operculum, opening by splitting; peristome lacking; elaters commonly produced with the spores

PLANT NAMES

All the kinds of organisms, including mosses, that have been discovered by biologists have been given scientific names. Such names are necessary in order for people to communicate with one another about organisms. The scientific names are also important in that they serve to indicate the relationships of organisms among one another. Although many kinds of plants have common names in everyday use—dandelion, live oak, pecan—most organisms do not have such a name. Thus, the scientific name is usually the only means we have of referring to most types of living things, and this is true for almost all mosses, few of which have useful common names.

Scientific names consist of two parts: the first part is the name of the genus (generic name) and the second part is the specific name. For example, *Quercus* is the name of the genus to which all true oak species belong. The specific name *virginiana*, when combined with the generic name *Quercus*, refers to the familiar Gulf Coast species of oak, the live oak. Other species of oaks include, for example, *Quercus alba*, white oak; *Quercus lyrata*, overcup oak; *Quercus nigra*, water oak; and *Quercus stellata*, post oak. Thus, closely related species are grouped together into the same genus.

In a similar fashion related genera are grouped together into higher categories called families. The common moss genera *Funaria* and *Physcomitrium*, plants which share many common features, are classified together into the family Funariaceae.

Botanists make up the names for the new genera and species of plants that they discover and describe, following a special set of rules. All scientific names are in Latin (or are Latinized—treated as though they were Latin words). No two kinds of plants may have the same name. The first letter of the generic name is capitalized, but the first letter of the species name is usually not capitalized. It is conventional to italicize scientific names in print and to underline them in writing.

By and large generic and specific names describe some feature or quality of the plant or commemorate a person, often one who had something to do with the genus or species. *Hedwigia*, a genus of mosses, was named in honor of Johannes Hedwig (1730–1799), author of an important book on mosses. Specific names often refer to the country or locality where the species was first discovered (*Syrrhopodon texanus*, the specific name referring to Texas, where the species was first found—at San Marcos). In formal botanical writing the name of the genus, when it is used alone, is followed by the name of the person, usually abbreviated, who first published a description of it (*Funaria* Hedw.). The two-part scientific name is followed by the name of the author who first published a description of the species—*Funaria hygrometrica* Hedw. When a species is reclassified from one genus to another, as has often happened, the name of the author who made the transfer is also written at the end of the scientific name, and the name of the original author is retained in parentheses—*Syrrhopodon parasiticus* (Brid.) Besch. When more than one author participates in publishing a description of a genus or species, the names of all of the authors are appended to the scientific name—*Papillaria nigrescens* (Hedw.) Jaeg. & Sauerb.

Relatively minor but consistent morphological deviations or variations from a typical species may be given formal taxonomic recognition as a variety—usually abbreviated as *var*. An example is *Ephemerum crassinervium* (Schwaegr.) Hampe var. *texanum* (Grout) Bryan & Anderson.

The system of scientific names may seem complicated and tedious to contend with, but it is a workable system, and there is no substitute for it at the present time. Although some of the names of mosses and other plants may seem to defy pronunciation at first glance (*Schwetschkeopsis*), patient sorting out of the syllables will generally lead to an acceptable pronunciation (*Schwetsch-ke-op-sis*).

FURTHER STUDY OF BRYOPHYTES

If your interest in the study of mosses has been whetted by this book, you may want to become more involved in bryology. You may also want to learn about liverworts. This section discusses some of the standard and specialized literature on the identification of mosses and liverworts and includes reference to the American Bryological and Lichenological Society, an organization devoted to the study of bryophytes and lichens.

Reference Works on Mosses

For many years the standard reference for the identification of North American mosses was the three-volume work *Moss Flora of North America North of Mexico*, published privately by Abel Joel Grout between 1928 and 1945. Although its coverage and nomenclature are now considerably dated, it is still a useful reference. Howard Crum and Lewis Anderson have recently (1981) written a comprehensive and beautifully executed two-volume technical treatment of the mosses of eastern North America, which provides coverage west to the plains states. A more popular work is Henry S. Conard's book *How To Know the Mosses and Liverworts*, which gives keys and illustrations for many North American mosses and liverworts.

The interesting assemblage of mosses in Florida has been treated by Ruth Schornherst Breen in *Mosses of Florida: An Illustrated Manual*, complete with keys, illustrations, and descriptions. "Mosses of the Interior Highlands of North America," by Paul L. Redfearn, Jr., in *Annals of the Missouri Botanical Garden*, 1972, treats the mosses of the Ozark and Ouachita uplands, which lie principally in Arkansas and Missouri. This work includes keys and descriptions but is not illustrated. It will be particularly useful to students in coastal plain areas adjacent to the highlands.

Reference Works on Liverworts and Hornworts

The most important reference for identification of liverworts and hornworts of eastern North America is *Hepaticae and Anthocerotae of North America East of the Hundredth Meridian* by Rudolf M. Schuster. This five-volume work is comprehensive, fully illustrated, and highly technical. Prior to the publication of Schuster's treatment, the standard reference for liverworts (and still the only one with coverage for all of North America) was *Hepaticae of North America*, by T. C. Frye and Lois Clark, published in five parts between 1937 and 1947. The book by Henry S. Conard, as mentioned above, includes the same level of treatment for liverworts and hornworts as it does for mosses.

The most useful publication for students of Gulf Coast liverworts and hornworts is a work by David Breil: "Liverworts of the Mid-Gulf Coastal Plain," published in the periodical *Bryologist* in 1970. This publication gives keys, descriptions, and illustrations of all liverworts and hornworts known to occur in the coastal plain area centering around Tallahassee, Florida. Although its utility decreases with distance away from the Tallahassee area, it is still a most useful reference in the area covered by the present book.

American Bryological and Lichenological Society

The American Bryological and Lichenological Society is the professional organization to which most North Americans interested in mosses belong, whether they are amateur or professional botanists. The society conducts annual meetings and publishes a quarterly journal, *Bryologist*, containing research articles on all aspects of mosses, liverworts, and lichens. Membership is open to all persons with an interest in these plants. For more information on membership and programs, consult a recent issue of *Bryologist* for the name and address of the current secretary-treasurer of the society.

LIFE STYLE AND STRUCTURE IN MOSSES

Life Cycle

Mosses are like virtually all other true plants in that they have two phases, or generations, in their life cycle. One phase includes the familiar green, leafy moss plant that, at maturity, produces the sex cells—eggs and sperm. In botanical terminology the leafy moss plant is called a gametophore (gamete, a sex cell; phore, to carry). The gametophore, together with the protonema, constitutes the gametophyte generation of mosses. When a moss sperm fertilizes a moss egg, the resulting zygote (fertilized egg) does not grow into another leafy moss plant. Instead, it matures into the other phase of the moss life cycle, the sporophyte. At maturity the sporophyte of most mosses consists of a "foot," connected to the gametophore, and a slender stalk, or seta, which bears at its upper end a capsule filled with spores. When the spores are released from the capsule they will, if they land in a suitable place, germinate and eventually produce new leafy moss plants. The spores of mosses, however, do not grow directly into the leafy moss gametophores; instead, they first produce a growth of delicate, green, algalike filaments called protonema (proto, first; nemat, thread). Buds arise from the protonemal filaments and grow into the leafy moss plants. In some mosses, especially those that grow on bare soil, the protonema may form conspicuous mats that can persist for a long time. But in most mosses the protonemal stage is brief and rather inconspicuous. Mosses and other bryophytes differ from all other land plants in having the gametophyte as the conspicuous dominant generation, with the sporophyte attached to it and more or less nutritionally dependent upon it. (Consult a good encyclopedia or textbook on botany for a more detailed account of the moss life cycle.) A glossary of technical terms commonly used in bryology is provided on page 237.

The Gametophyte

In many mosses the gametophores grow erect, anchored to the substrate by rootlike organs called rhizoids. Such mosses commonly bear their sporophytes at the tips of the stems and are thus said to be acrocarpous (acro, top; karpos, fruit). Most other mosses have the gametophores (or portions of them) prostrate, branched, and anchored to the substrate by rhizoids and bear their sporophytes along the stems and branches. Mosses of the latter type are said to be pleurocarpous (pleuro, side). Regardless of growth form, the gametophores (as well as the sporophytes) of most mosses are basically similar in structure.

Moss leaves vary widely among the different families of mosses and thus are extremely important in identification. Leaves may or may not have a costa, or midrib; the costa may run the length of the leaf or be shorter; it may be double, that is, there may be two costae per leaf. The margins of the leaves may be undifferentiated, that is, composed of cells similar to those in the rest of the leaf; or the marginal cells may be differentiated in various ways—

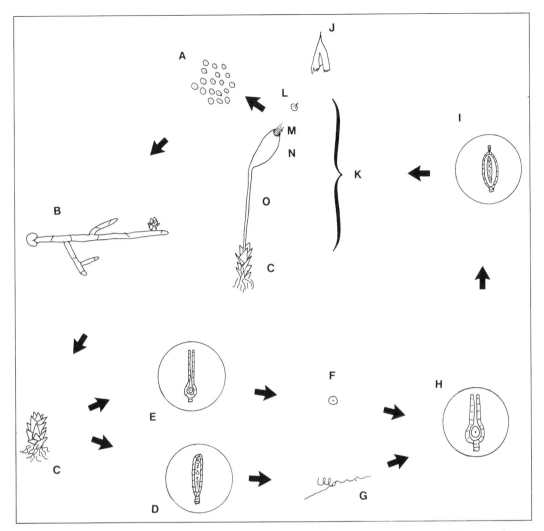

Fig. 1. Diagram of the life cycle of a moss. *A*, spores. *B*, protonema produced by a germinating spore; note developing leafy bud on the protonema. *C*, leafy moss plant—the gametophore. *D*, an antheridium produced on the gametophore; a single gametophore may produce numerous antheridia and each antheridium produces numerous biflagellate sperm. *E*, an archegonium produced on the gametophore; it contains an egg in the enlarged basal portion. Each gametophore may produce numerous archegonia, each forming a single egg. *F*, an egg. *G*, a single sperm—the body of the sperm is coiled. *H*, an archegonium containing a fertilized egg—the zygote. *I*, a modified, enlarging archegonium containing a multicellular embryo developed from the zygote. *J*, calyptra, developed from the archegonium and shed at maturity of the sporophyte. *K*, mature sporophyte; it is attached at the base of the seta by a foot, connecting it to the gametophore. *L*, operculum. *M*, peristome. *N*, capsule. *O*, seta. Features labeled *A–G* in the diagram represent the gametophyte generation in the moss life cycle. The sporophyte generation begins with the zygote.

color, shape, texture—or they may be in more than one layer. In many mosses the margins of the leaves have projections of various types, but in other mosses the leaf margins lack projections and are said to be entire. Also, the leaf margins may be plane, erect, inrolled or involute, or revolute. In the majority of mosses the leaf is more or less flat, but in some the leaves bear longitudinal grooves or folds and are then said to be plicate, and in others the leaves bear transverse ridges and depressions and are said to be undulate. Leaf shape is quite variable among mosses, ranging from nearly circular in outline to narrow and greatly elongate. There is also considerable variation in the cells of moss leaves, which may be essentially isodiametric or variously elongate. The arrangement of cells in the moss leaf may be prosenchymatous, in which the cells are elongate with their ends tapering and overlapping one another, or parenchymatous, in which the cells are more or less isodiametric with their ends flattened and abutting adjacent cells rather than overlapping. The exposed surfaces of the leaf cells may be flat and smooth, or variously bulging, or ornamented by small protrusions called papillae. Cells heavily beset with papillae are sometimes very obscure and difficult to distinguish under the microscope.

In most mosses the leaves are spirally arranged on the stem so that the stem, with its leaves, would be approximately circular in outline as seen in cross section. In many mosses, however, the leafy stems are flattened, or complanate. In a few mosses the leaves are borne in two opposite rows on the stem.

The sex organs in mosses are the antheridium, which produces sperm, and the archegonium, which produces the egg. A cluster of the cylindrical antheridia enclosed within usually modified leaves is called a perigonium; a cluster of the flask-shaped archegonia similarly enclosed is called a perichaetium. Many mosses bear both antheridia and archegonia on the same plant, but other mosses have some plants with only antheridia and others with only archegonia—thus there are male and female plants.

In addition to leaves and sex organs, stems and branches of mosses may bear small, often leaflike appendages, the paraphyllia and the pseudoparaphyllia, inserted among the leaves. Paraphyllia may be abundant and conspicuous, as in mosses of the genus *Thuidium*, or they may be small and scarce. Pseudoparaphyllia, if present, occur around the bases of branches and around branch buds.

Many mosses regularly bear brood bodies of various types that are specialized for vegetative reproduction. The types called gemmae are small, globular or elongated, and multicellular, and they may occur on the leaves, on the stem, in leaf axils, or on special, modified stem tips. Another type, called propagula, are leaves, buds, or modified branches that are specialized to fall and serve for vegetative reproduction. Some mosses form subterranean tubers on their rhizoids that may also serve for vegetative propagation.

The Sporophyte

A typical moss sporophyte consists of a foot, embedded in the gametophore, the seta, and the capsule. The upper part of the capsule is covered by a portion of the modified archegonium, now called the calyptra, which falls away when the capsule is fully mature. In most mosses there is a lid, the operculum, that comes off to allow release of the spores from the urn of the capsule. Some mosses have no definite operculum, in which case the capsules open irregularly to allow escape of the spores. An annulus, a ring of cells that aids in releasing the operculum, is often present in the capsule. In the majority of mosses the mouth of the urn, exposed by the fallen operculum, is surrounded by one or two circles of pointed, toothlike appendages, the peristome. Some mosses have a double peristome, some a single peristome, and some have none. The outer row of appendages (exostome) is composed of the teeth, with the same name applied to the appendages of mosses with a single peristome. The members of the inner circle of a peristome (endostome) are called segments. In some mosses slender structures called cilia alternate with the segments in the inner peristome. The peristome functions in regulating release of the spores from the urn. As a general rule peristomes open under conditions of low humidity and close under humid conditions, but the reverse is true in some mosses. The very base of the urn is often distinctly differentiated into the neck. In some mosses the neck may be enlarged, elongate, or otherwise modified, and if so may be called the hypophysis. The capsules of many mosses bear stomata, mostly near the base or on the neck. The stomata are similar to those found in higher plants and may or may not be functional in gas exchange.

While some mosses produce sporophytes routinely, many mosses have only rarely been found with sporophytes, and some are not known to produce sporophytes at all. Still others produce sporophytes regularly in one part of their range but only rarely or not at all in other parts of their range. It is conventional among many bryologists to refer to mosses bearing sporophytes as fruiting and to those lacking sporophytes as sterile. Mosses that fruit irregularly or not at all usually have highly effective ways of propagating the gametophyte generation asexually by means of specialized structures called brood bodies, referred to above.

Life Style and Ecology of Mosses

Most mosses are rather small; some are tiny, only a few millimeters tall. A few exotic mosses reach respectable sizes, up to a foot or more tall for individual stems. The generally small size of mosses is attributable to their lack of vascular tissue and roots, which forces them to remain small in order to take advantage of intermittently available surface water.

In spite of their small stature and seemingly delicate structure, mosses of some types can endure the rigors of subpolar regions of the world, and others can thrive under the harshness of desert climates. There are few places in the world where at least some mosses do not occur, although no mosses are truly marine.

Mosses are at home on virtually every imaginable substrate. Some are terrestrial, and some are aquatic; some grow on rocks, others on tree trunks, rotted logs, twigs, and leaves. Some types are even specialized for living on dung. In areas of high humidity, old walls, roofs, chimneys, and even asphalt shingles support various mosses. Their spores, or other forms of disseminules, are responsible for the invasion of suitable substrates by mosses.

Much speculation but little actual research has been expended on the effectiveness of moss spores in long-distance dispersal of mosses. The spores of most mosses are very small, and it would seem reasonable to believe that they would be carried great distances by air currents. This belief has a respectable history. Writing in the late 1700s, William Bartram noted in his delightful *Travels Through North & South Carolina* that "some other seeds, as of the Mosses and Fungi, are so very minute as to be invisible, light as atoms, and these mixing with the air, are wafted all over the world." In spite of the fact that the spores of many mosses are small enough to be wind transported, and even though the spores of many mosses can withstand extended periods of drying and freezing, it appears that actual instances of long-range dispersal by spores are quite rare. An excellent review of this and related subjects was published by Howard Crum (1972) in a paper on the geographic origins of the mosses of the deciduous forest of eastern North America.

Many mosses are perennial, living and growing from year to year. Others are annual in their life style, mostly appearing during late winter and early spring and then disappearing for the rest of the year, after producing spores. Such mosses persist in a given site as spores or other reproductive structures, such as tubers. Some mosses are highly specialized in their choice of substrate. *Anacamptodon splachnoides* grows almost exclusively in our area in and around seeping knotholes or cracks in living trees, and *Splachnobryum obtusum* occurs only on calcareous substrates such as limestone or old mortar work.

Many mosses form large, conspicuous masses when growing in favorable situations. *Sphagnum*, for example, commonly occurs in bushel quantities in prime sites. Other mosses may have an exceedingly scanty growth habit so that they occur as scattered individual plants or as clumps of a few plants. Unless one knows the proper habitat in which to search, it may be very difficult to find mosses of the latter types.

No mosses are known to be parasites of other plants. It is sometimes assumed that plants that grow upon other plants (such as a moss growing on a tree trunk) are parasites. True parasites, however, such as beechdrops and dodder, take some or all of their nutrition from their host, while mosses manufacture their own food and merely use the tree as a place to grow.

Mosses in general are modest plants occupying habitats in nature that are not otherwise exploited by the coarser vascular plants. In deciduous forests along the Gulf Coast mosses are most conspicuous during the winter months, after the leaves have fallen from the tree canopy. Then the stream banks are suddenly carpeted with green as mosses grow quickly in the wealth of sunlight. With spring, new leaves on the trees shade the moss carpets, dulling their vividness, and rank growths of spring herbs quickly hide the mossy banks.

COLLECTION AND CARE OF MOSSES

One of the most enjoyable aspects of the study of mosses (and of other plants as well) is that the student will always have a good excuse to go on field trips into natural areas to collect specimens for study. Collecting mosses, aside from such routine torments of the flesh as chiggers and mosquitos, is easy. The anticipation of discovering kinds of mosses that you have not seen before adds to the pleasure of collecting trips.

Mosses are probably the easiest of all plants to gather in the field and to prepare for filing in a collection after identification. This chapter tells you how to collect mosses, what is needed to do so, and how to make a reference collection, or herbarium, with the specimens that you identify.

Collecting Equipment and Techniques

Only simple equipment and materials are needed for collecting mosses; these are listed below.

> Pocket notebook
> Small paper bags
> Sturdy pocket knife
> Pencil or ballpoint pen
> Shoulder bag
> Hand lens or pocket magnifying glass

The notebook is useful for keeping a record of where and when you make collections and for notes on the habitats in which you collect. (Note-taking is a good habit to get into. Information not recorded at the time of a collection is likely to be imperfectly remembered later or lost altogether.) If you give a consecutive number to each specimen as you collect it, and also enter the numbers into the notebook along with basic information on each collecting site, then after you identify the specimens you can write their names by the proper number in your book. In this way you will have a complete record of what you have collected as well as where and when you collected it.

The small paper bags are used to hold the specimens as they are collected in the field until you can study them in detail later. The one-pound size, found in most grocery stores, is very convenient to use. Some people prefer to make collecting packets by folding stationery-size sheets of paper, but the paper bags are much more convenient. You can write field data directly onto the bags as you collect specimens. Do not use plastic bags for collecting mosses unless you plan to keep the moss in the living state. The intense humidity maintained in a plastic bag will cause the specimen to mold quickly, and the bag will complicate the drying process later when you prepare the specimens for storage.

The pocket knife is used for cutting mosses from their substrates—trees, logs, soil, rocks. If you buy a pocket knife with a metal loop at one end, tie a string through the loop, and hang the knife around your neck, it will be readily at hand. Buy brightly colored pencils or pens—they are easier to locate if dropped—for recording information on the paper specimen bags as you make each collection. The shoulder bag holds the specimens that you collect and other supplies, including extra pencils, collecting bags, lunch, a raincoat, and insect repellent.

The hand lens should have a magnifying power of at least 10x but not more than about 15x. Tie a string to it and hang it around your neck with the knife. It is used to take a close look at a specimen before you actually collect it, to make sure, for example, that the specimen is really a moss and to determine that the specimen is in good shape for collection. With practice you will be able to use the lens to identify in the field many of the mosses that you find. Hold the lens close to your eye with one hand; with the other hand bring the specimen into focus by slowly moving it close to the lens. Position yourself so that light can fall upon the specimen as you look at it through the lens. Tilt your head back so that the sunlight can fall directly on the specimen.

Step up to a growth of what you think is a moss. Lift a small portion of it from the substrate with the knife tip. Use your lens to verify that the specimen is a moss and that it is in suitable condition for collecting. Get out one of the small paper bags and write on it the type of substrate (tree trunk, sandstone, soil, log), the locality, and the general nature of the collecting site (pine or hardwood forest, old field), and then collect as much of the moss as you want and place it in the bag. Fold the top of the bag so the specimen will not fall out. Make sure that you still have your knife, pencil, and lens before you move on. A modest handful is ample for most collections. If you want to make duplicate specimens to send away later for someone else to identify for you, then you may want to take a little more than a modest handful. Some mosses never grow in handful quantities, and you must be content with what you can find. Mosses very often grow in mixtures of several different kinds.

Preparation and Maintenance of a Reference Collection

Many students find it interesting and useful to make permanent specimens of some or all of the mosses that they identify. In this way, not only is there a record (in the form of the specimen) of mosses identified, but the specimens can serve in the future as an aid in identification of further collections. Such a collection of plant specimens is called an herbarium. Among other functions, the herbarium serves to document the distributions of plants and provides information on fruiting times and substrate preferences. Many universities and other places where muscologists work maintain moss herbaria. The largest moss herbarium in the United States is in New York City, at the New York Botanical Garden, where more than 320,000 specimens of mosses from all over the world are kept for study. Harvard University, in its Farlow Herbarium of Cryptogamic Botany, has almost as many specimens of mosses as the New York Botanical Garden. The United States National Herbarium, at the Smithsonian Institution in Washington, D.C., also has a very large number of moss specimens. If you do not plan to keep a reference collection of mosses, then you will not need to read the rest of this chapter; go on to the next one. If you do intend to keep a collection, here is how to begin.

DRYING THE SPECIMENS. Mosses and other bryophytes are the easiest of all plants to prepare for placement in the herbarium. All that is required is thorough drying of the specimen, and this can be done by opening the paper bag in which you collected the moss, loosening the contents to permit air circulation, and then placing the bag on a table or windowsill for a few days until the specimen is dry. For faster drying, or in areas of high humidity, one can make a simple drying apparatus by constructing a wooden box with a wire screen partition across the middle. Install a light bulb receptacle below the screen. Place the opened bags containing the mosses on the screen, turn on the light, and heat from the bulb will dry the specimens. A box about 2 feet square and standing about 2 feet high is adequate for most purposes. In constructing the drying box, make sure that the bulb will not be so close to the floor, or to the specimens, as to cause danger of burning. An alternative method is to dry the specimens at low heat in the oven of the kitchen stove.

Moss specimens can also be pressed in a regular plant press, such as is used by collectors of vascular plants, and dried either with blotters or by heat. Pressing, however, usually distorts the natural form of the specimen. Air drying is the simplest way to preserve mosses and costs no energy.

SPECIMEN PACKETS   Make paper herbarium packets for your specimens as follows. Cut a piece of light cardboard so that it is 5½ inches wide and 4 inches high. This is the pattern around which to fold the packets. Place the pattern on a sheet of paper (standard business stationery size, 8½ × 11 inches) and fold the bottom edge of the paper up over the pattern so that the edge is flush with the top edge of the pattern. Crease the fold. Now fold the right and left margins of the paper equally towards the center of the pattern, against the edges of the pattern, and crease the folds. Finally, fold the top margin of the paper down towards the bottom and crease the fold. Remove the pattern and the packet is ready to use. Of course, you can make usable packets in other ways and in any size convenient for your purposes. The method and size described above are fairly standard for many moss herbaria, although there are numerous variations. Packets of the size described above fit conveniently into shoe boxes for filing and storage.

SPECIMEN LABELS   The scientific (and other) value of a specimen depends upon the quantity and quality of the information on the specimen's label. While having the name alone on a specimen may be adequate for some students of mosses, most would prefer also including information on some or all of the following points: date of collection, locality, substrate, the vegetation and nature of the collecting site, other bryophytes growing with the specimen, and any other information that might be useful or otherwise interesting. Information for making up the labels comes from data recorded in your notebook at the time of collection and from data recorded on the paper bag in which you collected the specimen. The label should

HERBARIUM
UNIVERSITY OF SOUTHWESTERN LOUISIANA (LAF)
PLANTS OF   TEXAS   NEWTON COUNTY

Fontinalis novae-angliae Sull.

Bay-gall shrub bog; 2 mi. E of Mayflower
Community, 1 mi. N of FM 255 along dirt
road; pine-oak-evergreen shrubs.

Attached to roots in a stream.

William D. Reese   12044   17 Nov 1973

Fig. 2. A typical label for an herbarium specimen.

also include your name as the collector and, if you keep a notebook, the number you assigned to the specimen in the notebook.

Some people write or type label information directly onto the front flap of the packet. Most find it more convenient to write or type label information onto small pieces of paper which can then be pasted on the front flap of the packet. It is not very expensive to have labels printed with basic information at a commercial print shop. An example is shown in Figure 2, with typical collection data typed on it.

Have the labels printed in sheets of several labels for convenience in typing and cut them out for attaching to the packets. Use a stationery-grade paper with at least 25% rag content for both labels and packets to avoid deterioration of the paper. Neither mucilage nor rubber cements last very long in an herbarium; instead, use any of the readily available liquid, white glues to attach labels to the packets.

CARE OF SPECIMENS   With the dried specimens placed in packets and filed in some fashion, little further care is necessary. Few insects will eat dried mosses as a rule, but in some cases tiny insects known as psocids, or booklice, will invade moss specimens and cause damage by eating soft parts of the plants. Psocids are almost invisible to the naked eye because they are so small and because they are virtually colorless. The first clue to their presence in a moss collection is often the detection of damaged specimens. By periodically sprinkling moth crystals into the boxes containing the specimens, it is usually possible to exclude psocids and other insects. Keep lids on the boxes.

FILING SYSTEMS   The simplest filing system for a small herbarium is alphabetical, by genus name and specific name. As the collection grows in size it will generally be easier to use if the specimens are filed alphabetically by family and then alphabetically by genus and species within each family. In large moss herbaria the specimens are often arranged in the sequence used in the publication *Die Natürlichen Pflanzenfamilien* (Brotherus 1924–25). This system groups related genera together and allows ease of comparison of specimens when attempting to make identifications.

USES OF THE HERBARIUM   The herbarium grows in utility as it grows in size. The specimens already identified are used to aid in identification of other specimens. Information on the specimen labels and the specimens themselves can be used to answer questions such as these: At what times of the year is a given kind of moss present in your area? When does it produce sporophytes? What kinds of substrates does it prefer? With what other plants does it grow? Is it common or rare in your area? How much variation do different populations of it show?

The American Bryological and Lichenological Society (see page 9) sponsors a Moss Exchange by means of which members can exchange duplicate identified specimens of their own collections for identified specimens from other collectors all over the world. In this way a student can acquire examples of mosses that do not occur in his area, as well as further examples of mosses that do occur there. It is easy to collect enough material of many mosses to make several specimens instead of just one. The extra specimens can then be exchanged for others.

IDENTIFICATION OF MOSSES

Identification of mosses is usually based on two levels of observation, with the first level involving features of the moss that can be seen with the naked eye (or with moderate magnification, as with the hand lens) and the second level involving features that can only be seen by using a compound microscope. Once one has learned the identity of a given species of moss, it is often possible, with the aid of a hand lens in some cases, to recognize the moss in the field thereafter with a fair degree of confidence.

A microscope equipped with lenses to give two levels of magnification, approximately 100x and 400x, is adequate for identification of mosses. A third lens, giving the microscope a lower level of magnification (scanning lens, approximately 25x total magnification in the microscope) is also very helpful. Glass microscope slides and cover slips, dissecting needles, and very fine-pointed forceps are needed to prepare moss specimens for study under the microscope. Single-edge razor blades are needed from time to time for cutting sections of leaves and stems. A binocular dissecting microscope, with magnifications from about 10x to 40x, is also very helpful in preparing dissections of mosses for study under the compound microscope.

It is easy to make a dissection of a moss for examination under the microscope. Put a drop of water on a microscope slide, place a stem or branch of the moss in the water and let it soak for a moment, and then scrape off lots of leaves with the needle and forceps. Drop a cover slip on the preparation, and you are ready to study the slide under the microscope. Specimens that are very dry may need to be immersed in hot water briefly in order to soften them prior to dissection; this is especially true for dry sporophytes. With a little practice, you can learn to heat specimens in water right on the microscope slide using paper matches—or better yet a small alcohol lamp—as the heat source. A small electric hotplate with a container of water that you can dip specimens into works very well.

Occasionally you may want to make a semipermanent slide of a moss you have dissected. This is easy to do using Hoyer's solution as the mounting medium (Anderson 1954). The ingredients for preparing the solution are:

Distilled water	50 cc
Gum arabic (U.S.P. flake)	30 gr
Chloral hydrate	200 gr
Glycerin	20 cc

Mix the ingredients in the above order at room temperature and shake or stir occasionally until the solution is clear. You can make smaller quantities by reducing the amounts of the ingredients proportionally. Keep the prepared solution in an airtight container; a medicine bottle with a rubber bulb dropper is fine. Mosses to be mounted in the Hoyer's solution must first be thoroughly soaked in water, using hot water if necessary to completely soak the plants. Blot the excess water from the moss and place the moss directly into a drop of the Hoyer's solution on a clean slide. After making the dissection, drop a cover slip onto the preparation and add, if necessary, enough Hoyer's solution to fill completely the space under the cover slip. After a day or so it may be necessary to add more Hoyer's solution to replace fluid lost by evaporation from under the cover slip. The resulting preparation will last more or less indefinitely. Chloral hydrate is a rather strong chemical, and you should take care to avoid prolonged contact with it on your hands. Wash your hands after handling Hoyer's solution.

While Hoyer's solution works well, it is rather tedious to mix up because the gum arabic takes a long time to go into solution. Furthermore, chloral hydrate is now a controlled substance and thus difficult to obtain. A new mixture for a semipermanent mounting medium for mosses has recently been proposed by Frahm (1981), which could prove to be superior to Hoyer's solution. Certainly it is easier to prepare. Frahm's recipe is a mixture of equal parts of mucilage, glycerine, and water. He uses commercially prepared mucilage, such as one buys at a stationery store for gluing paper. An additional advantage to Frahm's mixture is that it is safer to handle and to have around because it lacks the chloral hydrate.

In identifying mosses it is occasionally necessary to cut cross sections of leaves. As a rule it is easy to make cross sections of moss leaves, especially after a little practice. The most important requirement for satisfactory sections is a sharp, single-edge razor blade. (After a few uses the razor blade will dull and should then be replaced.) Always begin with well-soaked leaves and always work under the dissecting microscope.

If the moss to be sectioned has relatively large leaves, use forceps to remove five or six leaves and stack them, one on top of another, in a small drop of water on a microscope slide. Working under the dissecting microscope, hold the leaves down with one finger while you use the razor blade to make many thin slices across the stack of leaves. Push the slices off into the water as you cut them. Remove the uncut portions of the leaves from the slide, use forceps or needle to spread out the cut sections, and then drop a cover slip onto the preparation. Although many of the leaf sections prepared in this way will be too thick to be of much use, a few on each slide will be just right for viewing. If the moss has small leaves, use several branches piled up on one another for sectioning.

An alternative method for sectioning moss leaves, using mucilage, results in a greater number of high-quality sections but takes longer to accomplish. Place a drop of ordinary mucilage, such as is used for gluing paper, on a glass slide. Blot the soaked leaves or leafy branches to be sectioned, in order to remove excess water, and then dip them into the mucilage, stirring them around so that they are well coated with the mucilage. Now remove them from the mucilage and place them on a clean microscope slide, arranging and stacking them so that they will be in a good position for convenient sectioning. Set the slide aside to dry for an hour or so to allow the mucilage to become firm. When the mucilage-soaked leaves, now firmly attached to one another and to the slide, have achieved a consistency something like rubber, they can be easily sliced into very thin sections under the dissecting microscope. Cover the sections with water and wait a few minutes for the mucilage to dissolve. Then stir the sections around with a needle to disperse them, drop on a cover slip, and the preparation is ready for viewing under the compound microscope. Gentle warming will speed up the initial drying process.

Sometimes it is necessary or desirable to study a peristome under the compound microscope. Select a suitable capsule, either a mature one with the operculum still in place or one from which the operculum has fallen but which still has an intact peristome. If the capsule is dry, immerse it in boiling water briefly to soften it. Otherwise, place the capsule directly into a small drop of water on a microscope slide. Under the dissecting microscope, use needle or forceps to remove the operculum (if present), and then gently tap and press the capsule to force out most of the spores. Use a razor blade to cut the peristome free from the capsule, making the cut a little below the mouth of the capsule. Remove the remainder of the capsule from the slide. Then cut the peristome vertically into equal halves. Use forceps to arrange the halves so that one lies on the slide with the outer side of the peristome uppermost, and the other lies with the inner side up. Drop a cover slip on the preparation, and it is ready to view under the compound microscope. If you use too much water on the slide in making the preparation, the peristome will float around and will be difficult to handle.

Keys, such as the identification keys used in this book, enable the user to determine the name of an unknown specimen. As with most human endeavors, there is no such thing as a perfect key. Thus all keys must be used with appropriate caution and judgment. The keys herein present the user with two choices at each step; ideally, only one of the choices is correct for a given specimen. In the best case, the two choices at each step are mutually exclusive, but in practice it is not possible for all the pairs of choices to be totally contradictory. Qualifying phrases such as "plants usually" and "leaves mostly" creep into keys and are a warning to the user for appropriate wariness at that point. When you key a specimen out to what you believe to be its name, always turn to the description and illustrations for that name to see if they match your specimen. If the description and illustrations do not fit your specimen, or fit it only poorly, go back to the key and try again. In the case of a specimen that absolutely refuses to yield to the keying process, flip through the pages of illustrations looking for a match or for a similar moss that might give clues as to the proper relationship. Always start at the beginning of a key.

THE GULF COAST AS A HABITAT FOR MOSSES

Among the more important factors determining the occurrence and distribution of plants, including mosses, in a given area are climate, topography, and the history of the area. When mosses particularly are considered, the vegetation of the area is also an important factor.

The climate of the Gulf Coast is commonly described as subtropical, but this designation is most appropriate only for the lands closely bordering the Gulf of Mexico. Further inland the moderating effects of the warm waters of the Gulf diminish, and the climate becomes temperate, rather than subtropical.

In land areas adjacent to the Gulf the winters are mild, the summers warm, and the humidity is quite high most of the year. Seasonality is not as pronounced as in inland areas, and the changes from one season to another are blurred and gradual. Although killing frosts occur regularly throughout the area covered by this manual, freezing conditions are generally of short duration. Snow, ice, and sleet are rare. Winters are characterized by invasions, often quite rapid, of cold, dry air from the north, typically soon followed by an influx of warm, humid air from the Gulf. Afternoon thundershowers are typical along the central and eastern portions of the Gulf Coast in the summer.

Rainfall is abundant over much of the Gulf Coast but diminishes sharply west of the Sabine River. From the Florida Panhandle to the Sabine, the average annual precipitation (virtually all in the form of rain) is in the range of about 52 to 60 inches, with some restricted areas here and there averaging up to 66 inches annually. West of the Sabine River, however, average annual precipitation diminishes, reaching a low of about 18 to 24 inches in the lower Rio Grande valley. High humidity and precipitation favor the growth of mosses, providing that other factors of the environment are also appropriate.

The topography of the Gulf coastal plain is generally subdued, with most of the region relatively plane and low-lying. Elevations up to about 500 feet above mean sea level occur in inland areas in the region covered by this manual, but most of the land surface is much lower. Topographic prominences are expressed for the most part by bluffs along rivers; old, elevated erosion surfaces in inland areas; and a few emergent salt domes in Louisiana and Texas.

As a general rule, diversity of topography produces diversity of habitats for mosses. Thus areas along the Gulf Coast with varied topography have a more diverse moss flora than equivalent areas of less topographic diversity. High habitat diversity, coinciding with high annual precipitation in the eastern part of the Gulf Coast, results in a more diverse moss flora than in the western part, where there is lower annual precipitation and less topographic relief.

The term *vegetation*, as used by botanists, refers to the nature of the assemblage of plants growing in a given area, for example, forest, grassland, or scrub. (In contrast, the term *flora* refers to the specific kinds of plants growing in an area; for example, a forest may be composed of liveoak and hackberry—as well as many other species.)

The native vegetation of most of the Gulf Coast, from Florida to eastern Texas, is forest, including wooded swamplands as well as upland hardwood and pine forests. Freshwater, brackish, and saline marsh vegetation is important in some parts of the Gulf Coast, and prairie vegetation occurs in southwestern Louisiana, continuing on down the Texas coast.

Mosses are most abundant and diverse along the Gulf Coast in forested regions and least abundant and diverse in marshlands. As the forests diminish in importance west along the Gulf Coast, giving way to prairie vegetation and scrub, the abundance and diversity of the moss flora also diminish. With a reduction in diversity of habitat, there is a consequent reduction in diversity of mosses. Still, there are some mosses that typically grow in grassland vegetation and do not occur in forests.

The history of a given area, including its geologic history as well as its history of human influence, is also an important determinant affecting the kinds of mosses (and other plants) that occur in the area.

The geologic history is an important factor in determining how long the area has been available for occupation by plants; it also gives clues about the sources of the plants in the area, that is, where the plants came from to occupy the area. The latter point is very complex, but mosses of the Gulf Coast are mainly of eastern North American origin. To the basic eastern North American assemblage of mosses is added a number of mosses having tropical American affinities, and some that are indigenous to the coastal plain. A few have affinities with western North America, and some are chance introductions from other parts of the world, probably brought here inadvertantly through human activities.

Humans have been a powerful influence in modifying the natural vegetation of the Gulf Coast and thus influencing the occurrence and distribution of mosses, as well as of other types of plants (and of animals too). Removal of the original forest cover, and its replacement in extensive areas with agricultural lands or monoculture pine forest, has greatly affected the moss flora of the Gulf Coast. Stream channelization, impoundment of streams and rivers by dams, and alteration of natural drainage systems for more rapid runoff of precipitation have also been important in modifying the moss flora. Urbanization and industrial development have played a role too. It is possible, if not probable, that such human activities have resulted in some degree of modification of the climate along the Gulf Coast, which could also affect the occurrence and distribution of mosses and other plants.

It must not be overlooked that human activities, in addition to destroying or modifying natural environments, have at the same time increased certain types of habitats and even created new ones for mosses. For example, some kinds of mosses require a calcareous substrate, and where such substrates are lacking these mosses cannot occur. But calcicolous mosses do occur today in many areas along the Gulf Coast where natural calcareous substrates are lacking, utilizing man-made substrates such as tombstones, mortar work, and concrete structures. Furthermore, the potential habitat for some weedy mosses requiring disturbed sites in which to grow has been greatly increased through human activities such as agriculture and road construction. Certainly such mosses are far more abundant today than they were prior to the human-created expansion of suitable habitat for them. Without these artificially created and maintained disturbed sites, such mosses would have been restricted to naturally disturbed situations, such as actively eroding stream banks, sand bars, and the fresh soil exposed by overturned trees.

Part Two

MANUAL OF THE
MOSSES OF THE
GULF SOUTH

The sequence of families and genera of mosses in this work is basically that of Brotherus (1924–1925). Nomenclature follows, for the most part, that used by Crum and Anderson (1981). Plants, for various reasons, have often been given more than one scientific name. Such duplicate names are called synonyms, and they are given in this book only where the names used here differ from those used by Breen (1963) or from those used by Crum and Anderson (1981). Synonyms, where used, appear immediately below the accepted names.

The family treatments in the manual portion of this book are organized as follows: family name (names of plant families end with *-aceae*); formal botanical description of the family and general comments; key to the genera (if the family includes more than one genus); the genus name; botanical description of the genus (if there are two or more genera in the family and if the genera include two or more species); general comments on the genus; key to the species (if the genus includes more than one species); botanical description of the species; notes on habitat, total range, and Gulf Coast distribution, and comments on differentiation from similar species. Comments on natural history or other interesting aspects of particular species are included from time to time, and notes on erroneous or unverified reports of other taxa from the area covered by the manual are included at the end of the generic treatments as appropriate.

The keys in this book enable the user to classify mosses correctly. These keys are composed of numbered pairs (couplets) of mutually contradictory choices, with both choices in a given pair bearing the same number. In using the keys, always start with the first couplet, that is, the pair of choices bearing the number 1. Read the two choices carefully and, by studying the specimen at hand, determine whether the first choice or the second describes the specimen most accurately. After deciding which choice best fits your specimen, follow the last line of the choice over to the right-hand column, where you will find either a name (with a page number) or just a number. If there is a name, it will give the identity of your specimen at the family, genus, or species level; the accompanying page number directs you to the description in the text. If there is only a number at the right, the number refers to the next couplet that must be used in the key. Go directly to that couplet, even if you must skip over intervening couplets to do so. For example, if you are at couplet number 3 and pick the second choice, go directly to couplet number 6, skipping over couplets 4 and 5. It should be noted, as a general caveat, that sterile mosses (those lacking sporophytes) may be very difficult, or impossible, to identify satisfactorily.

Key to Mosses of the Gulf South

1. Leaf cells of two distinct types, some large, empty, and porose, these surrounded by a network of smaller, narrow, green cells; leaves one cell layer thick, ecostate	SPHAGNUM p.35
1. Leaf cells essentially all alike, or if of more than one type, then not forming a network; costa present or lacking	2
2. Leaves clearly vertically inserted on stems, in two rows on opposite sides of stem, blade doubled adaxially at base of leaf and clasping stem	FISSIDENS p.51
2. Leaves inserted horizontally, in more than two rows, blades not doubled at base	3
3. Upper surface of costa bearing filaments or lamellae	4
3. Upper surface of costa lacking filaments and lamellae	6
4. Upper surface of costa bearing filaments	Pottiaceae p.96
4. Upper surface of costa bearing lamellae	5
5. Plants tiny; costa excurrent into a pale hair-point	PTERYGONEURUM OVATUM p.114
5. Plants small to robust, often coarse; leaves lacking hair-points	Polytrichaceae p.239
6. Plants pale, whitish green; leaves several cell layers thick, in cross section showing rows of small green cells and large empty cells	7
6. Plants shades of green; leaves mostly one cell layer thick except for the costa, margins sometimes thickened	8
7. Upper part of leaf flat, ligulate	OCTOBLEPHARUM ALBIDUM p.96
7. Upper part of leaf channeled, acuminate	LEUCOBRYUM p.90
8. Leaves with two distinct costae extending to midleaf or beyond	Hookeriaceae p.180
8. Leaves with costa single or lacking, or costa double and short, at base of leaf only	9
9. Plants small to tiny; stems erect; capsules immersed to barely exserted	10
9. Plants small to large, stems erect to prostrate; capsules exserted, or if immersed then at least primary stems prostrate	22
10. Capsules lacking an operculum, opening irregularly	11
10. Capsules with an operculum	19
11. Plants silvery green; stems erect and arising from a fleshy subterranean rhizome; spores very large, to ca 190 μm diameter	LORENTZIELLA IMBRICATA p.122
11. Plants various shades of green but not silvery; rhizome lacking; spores mostly smaller (to ca 300 μm in *Archidium*)	12
12. Capsules with a conspicuous neck	BRUCHIA p.78

12. Capsules lacking a conspicuous neck	13
13. Leaf margins strongly involute, wet and dry	ASTOMUM p.99
13. Leaf margins plane to revolute	14
14. Plants bulbiform, sometimes 3-angled; leaves broad and deeply concave	ACAULON p.113
14. Plants not bulbiform or 3-angled	15
15. Plants tiny, scarcely visible to the unaided eye, arising from a conspicuous, persistent protonema	Ephemeraceae p.122
15. Plants larger, easily visible to the unaided eye; protonema scanty or not apparent	16
16. Spores large, ca 50–300 μm diameter	ARCHIDIUM p.69
16. Spores smaller, not reaching 50 μm	17
17. Leaf margins recurved, at least at midleaf; cells mostly papillose	PHASCUM p.113
17. Leaf margins plane; cells smooth	18
18. Leaves lanceolate to ovate, toothed above	PHYSCOMITRELLA p.127
18. Leaves subulate, entire above	PLEURIDIUM p.73
19. Plants corticolous, dark green; peristome present	ORTHOTRICHUM p.159
19. Plants terrestrial, color various; peristome present or lacking	20
20. Perichaetial leaves setaceous; sporophyte highly oblique	DIPHYSCIUM p.231
20. Perichaetial leaves broader; sporophyte symmetric	21
21. Plants growing on rock, mostly dark green; leaves often with hyaline awns; cells small and dense; peristome present	GRIMMIA p.120
21. Plants growing on soil, green to yellowish green; leaves lacking hyaline awns; cells lax; peristome lacking	Funariaceae in part p.126
22. Stems erect; sporophytes mostly terminal; capsules exserted	23
22. Stems (at least the primary stems) prostrate (primary stems sometimes rhizomelike and inconspicuous); sporophytes arising along the stems, or on their branches, or both; capsules mostly exserted, sometimes immersed	55
23. Capsules with long, slender neck longer than the urn; plants terrestrial in disturbed sites	TREMATODON p.81
23. Capsules lacking a neck, or neck present but shorter than urn; habitat various	24
24. Leaf margins conspicuously thickened, or bordered at least in part with elongate, narrow, hyaline cells; lower portion of leaf enclosing large, conspicuous, clearly demarcated areas of hyaline cells sharply set off from green cells above and extending from costa nearly to leaf margins; green cells mostly papillose	SYRRHOPODON p.92
24. Leaves lacking conspicuous areas of hyaline cells, or if hyaline cells present then leaf margins not thickened or not bordered with narrow hyaline cells	25
25. Leaf cells papillose	26
25. Leaf cells smooth or bulging, but not papillose	29
26. Leaf cells unipapillose, the papilla either central or at end of the cell	27
26. Leaf cells pluripapillose	Pottiaceae in part p.96
27. Plants corticolous	PTYCHOMITRIUM p.156
27. Plants terrestrial	28
28. Papillae at ends of cells	Bartramiaceae p.150
28. Papillae central on cells	AULACOMNIUM p.149

29. Plants corticolous, growing well up on tree bark — 30
29. Plants terrestrial or on rocks, sometimes on tree bases, or logs and rotting wood — 32
    30. Alar cells differentiated, brownish; slender flagelliform branches almost always present in upper leaf axils — DICRANUM FLAGELLARE p.89
    30. Alar cells not differentiated; flagelliform branches lacking — 31
31. Costa excurrent into an awn; calyptra cucullate — BRACHYMENIUM p.137
31. Costa percurrent; calyptra mitrate — PTYCHOMITRIUM DRUMMONDII p.156

    32. Alar cells conspicuously differentiated, at least on some leaves, enlarged and often colored — Dicranaceae p.77
    32. Alar cells not particularly differentiated — 33
33. Leaves distinctly bordered with narrow, elongate cells — 34
33. Leaves lacking a distinct border, or border present but weak and indistinct — 35
    34. Leaf cells elongate; margins mostly relatively weakly bordered and entire to serrate; prostrate stems lacking — BRYUM p.139
    34. Leaf cells ± isodiametric or slightly elongate; margins mostly strongly bordered and bearing stout teeth; prostrate sterile stems often present — MNIUM p.145
35. Margins prominently toothed with sharp, paired teeth — RHIZOGONIUM SPINIFORME p.148
35. Margins entire to serrate or dentate; teeth, if present, single — 36
    36. Leaf cells prominently bulging, at least ventrally; plants tiny, strict calciphiles — 37
    36. Leaf cells flat to somewhat bulging; size and habitat of plants various — 38
37. Basal cells conspicuously enlarged and hyaline, forming a V-shaped area — LUISIERELLA BARBULA p.119
37. Basal cells not forming a hyaline V-shaped area — BARBULA AGRARIA p.109
    38. Leaf apex broadly rounded; leaves broad, oblong to spatulate — Pottiaceae in part p.96
    38. Leaf apex acute to acuminate; leaves narrow to broad, mostly variously lanceolate — 39
39. Upper and median cells ± isodiametric, mostly rather thick-walled and dense — 40
39. Upper and median cells elongate, often lax — 43
    40. Stem tips commonly elongated into naked, setalike extensions that usually bear brood bodies; leaf margins recurved in lower ¾ — AULACOMNIUM PALUSTRE p.149
    40. Stem tips leafy, lacking (or only rarely with) setalike extensions; leaf margins plane or only recurved near leaf base — 41
41. Plants tiny, delicate, growing in disturbed sites (parks, lawns); terminal cell of leaf sharp, usually reflexed and brown — TORTULA RHIZOPHYLLA p.117
41. Plants small to robust, scarcely delicate, mostly growing in natural habitats — 42
    42. Growing on rocks; leaf margins irregularly bistratose above — PTYCHOMITRIUM in part p.156
    42. Growing on soil; leaf margins unistratose — AULACOMNIUM in part p.149
43. Plants usually in wet or damp sites, glaucous green, water repellent when dry; stems often bearing clusters of brood bodies at the tips — PHILONOTIS in part p.151

Key to Mosses of the Gulf South (continued)

43. Plants of drier sites, green to yellowish green, readily soaking
    up water when dry; brood bodies lacking — 44
   44. Plants with sporophytes (sporophytes sometimes im-
     mersed and inconspicuous) — 45
   44. Plants lacking sporophytes. NOTE: It is often difficult to
     make satisfactory determinations of some of the follow-
     ing mosses when they are sterile; compare specimens
     with illustrations and descriptions — 51
45. Peristome lacking — PHYSCOMITRIUM p.128
45. Peristome present — 46
   46. Peristome double (segments of endostome sometimes short
     and inconspicuous, hidden by the teeth unless viewed
     from inside) — 47
   46. Peristome single — 48
47. Calyptra large, inflated at base, long-beaked — FUNARIA p.130
47. Calyptra small, not inflated — Bryaceae p.135
   48. Peristome teeth associated in 2's or 4's, reflexed
     when dry — SPLACHNUM PENNSYLVANICUM p.134
   48. Peristome teeth standing free from one another, erect or ±
     spreading when dry but not reflexed — 49
49. Calyptra inflated at base, with long beak — ENTOSTHODON DRUMMONDII p.133
49. Calyptra small, not inflated — 50
   50. Peristome teeth split to base into narrow divisions — DITRICHUM p.76
   50. Peristome teeth split only to about middle — DICRANELLA p.82
51. Leaves, at least upper ones, narrow, often acuminate-subulate — 52
51. Leaves broader, acute but never subulate — 54
   52. Costa long-excurrent and forming the subula; leaves never
     secund — 53
   52. Costa ending below leaf apex to shortly excurrent; leaves
     often distinctly secund — DICRANELLA p.82; DITRICHUM p.76

53. Upper margins often coarsely ciliate-toothed; plants matted
    with red rhizoids below, commonly growing on dung or
    other enriched substrates; rhizoid tubers lacking — SPLACHNUM PENNSYLVANICUM p.134
53. Upper margins serrate but not ciliate-toothed; rhizoids purple
    to brownish, not particularly matted; plants not preferring en-
    riched substrates; brown to purple tubers produced on the
    rhizoids and sometimes in the lower leaf axils — LEPTOBRYUM PYRIFORME p.138
   54. Leaves soft; cells large and lax; rhizoid tubers lacking — Funariaceae p.126
   54. Leaves firm; cells smaller and with firm walls; rhizoid tu-
     bers often present — Bryaceae p.135
55. Leaf cells uni- or pluripapillose (taxa variable in this regard
    are keyed both here and below) — 56
55. Leaf cells smooth to bulging, not evidently papillose — 66
   56. Leaves ecostate, or costa short and double and confined to
     base of leaf — 57
   56. Leaves with costa reaching midleaf or further — 60

57. Leaf apices hyaline — HEDWIGIA CILIATA p.169
57. Leaf apices concolorous — 58
    58. Leaves falcate-secund; papillae at upper ends of cells — CTENIDIUM MOLLUSCUM p.229
    58. Leaves straight; papillae over cell lumens — 59
59. Cells densely pluripapillose; leaves mostly complanate — ERPODIUM p.154
59. Cells unipapillose; leaves not complanate — SCHWETSCHKEOPSIS FABRONIA p.186

    60. Stems bearing conspicuous paraphyllia among the leaves; plants mostly regularly 1–3 pinnately branched — Leskeaceae in part p.187; Thuidiaceae p.197

    60. Stems lacking paraphyllia, or paraphyllia scarce and inconspicuous; plants mostly irregularly branched — 61
61. Plants growing on soil, rocks, tree bases, logs and stumps, not occurring well up on trees and shrubs — 62
61. Plants epiphytic, growing well up on stems and branches of trees and shrubs — 63
    62. Leaves with large, conspicuous areas of quadrate cells in alar regions — STEREOPHYLLUM RADICULOSUM p.217

    62. Leaves lacking well-defined areas of quadrate cells — Leskeaceae in part p.187; Thuidiaceae in part p.197

63. Median and upper cells serially papillose, each cell bearing a distinct row of small, sharp papillae — Meteoriaceae p.176
63. Median and upper cells variously papillose but the papillae not in rows — 64
    64. Alar cells quadrate in conspicuous groups — LEUCODONTOPSIS GENICULATA p.174

    64. Alar cells undifferentiated or vertically elongate, nor particularly quadrate — 65
65. Cells in leaf base vertically elongate, with thickened sinuose-pitted walls, very different from median and upper cells; calyptra campanulate, hairy — MACROMITRIUM RICHARDII p.160
65. Cells in leaf base not much different from median cells; calyptra cucullate, naked — Leskeaceae in part p.187
    66. Costa single, reaching midleaf or further (taxa variable in this regard are keyed both here and below) — 67
    66. Costa lacking or short and double and confined to base of leaf (taxa variable in this respect are keyed both here and above) — 89
67. Plants aquatic or growing in wet places (habitat may be exposed and dry at times due to falling water level) — 68
67. Plants of varied habitats but not characteristically growing in wet sites — 72
    68. Plants slender; leaves very narrow, appearing setaceous because of long-excurrent costa, often secund at branch tips — DICHELYMA CAPILLACEUM p.167
    68. Plants slender to robust; leaves lanceolate or broader — 69
69. Plants erect from creeping rhizomes, often ± dendroid; leaf tips coarsely toothed; paraphyllia abundant — CLIMACIUM AMERICANUM p.168
69. Plants prostrate or floating, lacking rhizomes and paraphyllia,

or paraphyllia present but small and inconspicuous; leaf tips
entire to serrate — 70

    70. Leaves ± distinctly bordered by elongate cells; costa
percurrent — 71

    70. Leaves lacking distinct border; costa clearly ending below
leaf tip — AMBLYSTEGIUM p.207

71. Plants growing in prostrate mats attached to rocks in fast-
flowing streams; leaf border very prominent, of yellowish,
thick-walled cells — SCIAROMIUM LESCURII p.206

71. Plants growing in floating masses attached to various substrates
in still waters or slow streams; leaf border sometimes indistinct, lacking yellowish, thick-walled cells — BRACHELYMA SUBULATUM p.167

    72. Plants corticolous, with creeping, rhizomelike primary
stems (often inconspicuous) and erect (or pendent)
branched or simple secondary stems; primary and secondary stems mostly quite different from one another — 73

    72. Plants of varied habitats; if corticolous then lacking a dif-
ferentiated rhizomelike creeping primary stem; stems and
branches essentially all alike — 80

73. Secondary stems and branches strongly flattened — NECKEROPSIS UNDULATA p.179

73. Secondary stems and branches terete, or at least not strongly
flattened — 74

    74. Leaves either crispate or spirally twisted around stems
when dry; secondary stems mostly simple and very short — Orthotrichaceae in part p.158

    74. Leaves straight when dry or slightly curved, neither crispate
nor twisted around stems; secondary stems mostly elongate (short in *Homalotheciella*), simple or branched — 75

75. Secondary stems simple (occasionally with few, irregular
branches) — 76

75. Secondary stems freely subpinnately branched — 79

    76. Secondary stems very short, crowded in two rows along
opposite sides of creeping primary stems; plants very
small, slender and soft — HOMALOTHECIELLA SUBCAPILLATA p.209

    76. Secondary stems elongate, not crowded in two rows; plants
medium-size — 77

77. Costa serpentine above; secondary stems usually curved or
coiled at tips when dry; leaf tips mostly coarsely toothed — HERPETINEURON p.197

77. Costa straight; secondary stems straight; leaf margins entire — 78

    78. Secondary stems very slender, tightly foliated with imbri-
cate leaves when dry; axillary gemmae lacking; sporophytes commonly present, immersed — CRYPHAEA p.170

    78. Secondary stems ± turgid, loosely foliated when dry, the
leaves somewhat spreading; uniseriate gemmae usually
present in leaf axils; not producing sporophytes in
our area — JAEGERINA p.175

79. Axillary, short, uniseriate gemmae usually present; leaves very
concave, loose and somewhat spreading when dry, not appearing plicate; sporophytes not produced in our area — PIREELLA p.175

79. Axillary gemmae lacking; leaves not particularly concave,

mostly imbricate and appressed when dry, often appearing plicate; sporophytes usually present — FORSSTROEMIA TRICHOMITRIA p.172

80. Plants very small, soft and delicate, commonly growing on trees, sometimes on rocks, logs, and stumps, mostly ± glossy; capsules erect and symmetric — 81

80. Plants mostly medium-size to robust, mostly not corticolous; glossy or dull; capsules various — 82

81. Plants yellowish green; stems creeping, closely subpinnately branched, the branches erect or curved and very short; leaves narrow, abruptly slenderly acuminate; median leaf cells linear-fusiform; seta papillose; calyptra with delicate hairs — HOMALOTHECIELLA SUBCAPILLATA p.209

81. Plants dark green to brownish; stems freely and mostly irregularly branched; leaves acute to broadly acuminate, sometimes slenderly acuminate; median cells somewhat elongate; seta smooth; calyptra naked — Fabroniaceae in part p.183

82. Alar cells quadrate in large, conspicuous areas; plants often flattened — STEREOPHYLLUM RADICULOSUM p.217

82. Alar cells not quadrate, or if quadrate then few and not forming large, conspicuous areas; plants mostly not or only slightly flattened — 83

83. Plants small, dark green or brownish, dull; median cells isodiametric, firm, obscure; capsules erect and symmetric or nearly so — LESKEA in part p.192

83. Plants small to robust, green to yellowish green, often glossy; median cells elongate; capsules mostly curved and asymmetric — 84

84. Costa terminating in a distinct tooth dorsally; terminal cells of leaf differentiated, very short — EURHYNCHIUM p.214

84. Costa mostly lacking a distinct dorsal tooth (or tooth, if present, very small); terminal cells of leaf short or elongate — 85

85. Stems and branches very turgid; leaves deeply concave, tips apiculate, often twisted — BRYOANDERSONIA ILLECEBRA p.212

85. Stems and branches not particularly turgid; leaves not or only somewhat concave, tips variously acuminate but not apiculate, mostly plane (twisted in *Rhynchostegium*) — 86

86. Leaves distinctly squarrose — CAMPYLIUM in part p.204

86. Leaves erect to spreading but not squarrose — 87

87. Leaf tips slenderly acuminate and twisted; leaves not plicate, margins serrate nearly all around; stems usually flattened; costa often terminating in a very small tooth dorsally — RHYNCHOSTEGIUM SERRULATUM p.213

87. Leaf tips acute to acuminate, not twisted; leaves plicate to plane, margins serrate to entire; stems various, often flattened; costa lacking a dorsal tooth — 88

Key to Mosses of the Gulf South (continued)

88. Leaves mostly distinctly plicate; margins serrate; stems ± terete, not particularly flattened — BRACHYTHECIUM p.210

88. Leaves not plicate; margins mostly entire, sometimes slightly serrulate; stems often flattened — Amblystegiaceae in part p.203

89. Plants aquatic (sometimes exposed by falling water level) — FONTINALIS p.164

89. Plants growing in various habitats, but not in water — 90

    90. Primary stems prostrate, creeping, rhizomelike, quite different from secondary stems (similar but more slender in *Leucodon*), sometimes inconspicuous; secondary stems erect (or pendent), simple or branched; plants corticolous, sometimes on logs and stumps, rarely rocks — 91

    90. Primary stems not differentiated or not rhizomelike, if prostrate and creeping then essentially similar to secondary stems and branches; plants of various habitats — 93

91. Secondary stems freely subpinnately branched — FORSSTROEMIA TRICHOMITRIA p.172

91. Secondary stems simple or irregularly few-branched — 92

    92. Secondary stems usually crowded, often curved, their leaves tightly imbricate when dry; axillary gemmae lacking; sporophytes commonly present — LEUCODON JULACEUS p.173

    92. Secondary stems usually remote, straight, their leaves loosely erect-spreading when dry; uniseriate axillary gemmae usually present; sporophytes not produced in our area — JAEGERINA SCARIOSA p.175

93. Plants slender; leaves squarrose — CAMPYLIUM in part p.204

93. Plants slender to robust; leaves straight to falcate-secund but not squarrose — 94

    94. Alar cells quadrate in conspicuous groups, extending up along lower margins — 95

    94. Alar cells various but mostly not quadrate, if some quadrate then not forming conspicuous areas that extend up along the margins — 99

95. Plants dark green to brownish green, glossy, with clusters of leafy brood bodies in axils of upper leaves on at least some of the erect branches — PLATYGYRIUM p.222

95. Plants dark to yellowish green; leafy brood bodies lacking — 96

    96. Leaves strongly falcate-secund — HYPNUM IMPONENS p.224

    96. Leaves various but not falcate-secund — 97

97. Leaves tightly imbricate; stems and branches all prostrate, usually ± flattened — ENTODON p.215

97. Leaves various but not tightly imbricate; branches mostly erect or ascending, not particularly flattened — 98

    98. Leaves serrate above, sometimes all around; seta papillose; calyptra with slender hairs — HOMALOTHECIELLA SUBCAPILLATA p.209

    98. Leaves entire or nearly so; seta smooth; calyptra naked — Hypnaceae in part p.221

99. Stems turgid; leaves deeply concave, apiculate, the apiculation often twisted — BRYOANDERSONIA ILLECEBRA p.212

99. Stems not particularly turgid; leaves mostly ± plane, not out-

standingly concave, acute to acuminate, apex neither apiculate nor twisted   100

  100. Plants very small, slender; leaves to 0.5 mm long   PLATYDICTYA p.223

  100. Plants small to robust; leaves larger   101

101. Leaves complanate, straight; alar cells not or only slightly
differentiated   Hypnaceae in part p.221

101. Leaves mostly not complanate, or if somewhat complanate
then falcate-secund; alar cells mostly conspicuously inflated
(quadrate in *Hypnum imponens*)   102

  102. Plants small; leaves straight, ascending to homomallous;
median cells rhombic to linear   SEMATOPHYLLUM p.218

  102. Plants medium-size; leaves falcate-secund, at least at
stem and branch tips, sometimes complanate; median
cells linear   103

103. Plants conspicuously golden-glossy; leaves complanate; cells
yellow across base of leaf   BROTHERELLA p.218

103. Plants not particularly golden or glossy; leaves not complanate
or only somewhat so; cells not yellow at leaf base   HYPNUM in part p.224

SPHAGNACEAE Dum.

Mostly soft, pale green or glaucous plants growing in usually dense
clumps or cushions or floating, occasionally dark green, brownish,
blackish, or tinged with pink or red. Stems mostly organized into
nodes and internodes, the internodes bearing spirally inserted
leaves, the nodes bearing one-sided clusters of leafy branches,
some of which hang down and are appressed to the stem, the others ± divergent; stems and branches mostly with an outer layer of
hyaline cells (cortical cells). Leaves ecostate, composed mainly of
two types of cells, some small, linear, and green, the others enlarged, hyaline, empty, porose, and mostly with spiral thickenings
on the walls (hyaline cells), the green cells in sinuous interlocking
rows enclosing the hyaline cells; leaves mostly bordered by one or
more rows of narrow, hyaline, elongate cells, the outermost row of
which is often partially eroded forming the resorption furrow. Sporophytes lacking setae, the capsule sitting directly upon the foot and
both elevated at maturity upon a setalike extension of the fertile
gametophore branch, the pseudopodium; calyptra delicate, tearing
away irregularly at maturity; peristome absent; operculum explo-
sively detached at maturity and thrown off with the spores. Not often fruiting in our area.

    The family includes only the single genus, *Sphagnum*, of world-
wide occurrence.

Sphagnum is one of the few genera of mosses with a generally accepted and widely used common name—the peat or sphagnum moss. These mosses grow, as a rule, in wet sites. In the northern part of North America, peat mosses commonly grow in bogs and form a major portion of the vegetation in such sites. In the Gulf coastal plain, however, peat mosses only rarely form important proportions of local vegetation, doubtless due to the lack of suitable bog areas. However, conspicuous populations of several species of *Sphagnum* can sometimes be found in especially favorable habitats in the coastal plain, especially east of the Mississippi River.

Common habitats for peat mosses in the coastal plain include wet pine forests and savannahs on sandy soils, forest springs and stream banks, seeping sandstone outcrops, and ravine bottoms. There is a fairly good correlation between the occurrence of native pine vegetation and peat mosses in the coastal plain. Where native pines are absent, peat mosses are likely to be absent also. Thus the coastal prairies of Louisiana and Texas are as devoid of peat mosses as they are of native pines.

Peat mosses are the most important of all mosses from the commercial point of view. Because they render their habitat acid, which retards or suppresses decay, their remains and those of other plants growing with them tend to accumulate in northern bogs, forming the harvestable product sold as peat. The dried moss itself is also harvested in quantity and sold as peat moss. Both products are familiar to gardeners and greenhouse keepers for their useful properties as soil conditioners. Sphagnum moss is also justly renowned for its water absorption and retention quality and was used widely in former times as diaper material for infants and as bandaging for wounds.

Sphagnum has traditionally been regarded by bryologists as a difficult genus, meaning that it was thought to be difficult to identify and recognize species in it. While it may be true that, at least at first glance, many species of *Sphagnum* do depressingly resemble one another in the field, the species can be keyed out satisfactorily, and many of them can be recognized in the field with experience. The long history of difficulty in determining *Sphagnum* species was noted in 1880 by the British muscologist Braithwaite, who spoke of *Sphagnum* as having "exercised the botanist by the difficulty which attends their correct determination."

Although nomenclature for the majority of mosses dates back to Hedwig (1801), that for *Sphagnum* begins with Linnaeus in his *Species Plantarum* (1753), where he classified essentially all *Sphagnum* under a single species, *Sphagnum palustre*.

Special Study Methods for *Sphagnum*
Identification of peat mosses in the following key is based on features observable under the microscope on both the inner and outer surfaces of the branch leaves, on cross sections of the branch

leaves, and on the cortical cells of the branches and sometimes the stems. The procedure for preparing a microscope mount of *Sphagnum* for identification is not difficult and takes only a few minutes. From a soaked *Sphagnum* stem select four or five well-developed leafy branches of the divergent type from near the tip of the stem and place them in water on a glass slide. Working under the dissecting microscope, make a stack of three of the branches, all pointing the same way, and hold the stack down with your finger on the bases of the branches. Use a razor blade with your other hand to cut many cross sections of the branches and their leaves, pushing the sections out of the way as you cut them. Remove the uncut basal portions of the branches and discard them. Now use forceps to remove the leaves from the other branches on the slide, taking care to have some leaves with the inner surface up and some with the outer surface up. Leave the defoliated branches on the slide. Spread the leaves, leaf sections, and defoliated branches evenly on the slide and drop a cover slip onto the preparation. It is now ready to study under the compound microscope. A cross section of the stem may also be necessary for identification.

The sections of the leaves will remain curved, following the original shape of the uncut leaf, so that it is easy to tell which is the inner surface (that originally next to the stem), and which is the outer surface (that originally away from the stem). It is also easy to see whether or not a resorption furrow is present by looking at the edges of the leaf sections. The furrows, if present, will show up as U-shaped depressions in the small cells at the very ends of the sections.

In the following key, all references to green cells, hyaline cells and resorption furrows are based on observations of leaves of the divergent branches as seen in cross section.

1. Stems mostly unbranched or branches scattered and single, not in typical clusters at nodes; plants turgid, ± julaceus	1. S. CYCLOPHYLLUM
1. Stems bearing branches in clusters at nodes, the clusters of branches sometimes crowded on the stem	2
2. Hyaline cells lacking spiral thickenings, with a row of pores down the middle, or with two rows of pores; plants coarse and appearing bristly when dry, lacking the typical *Sphagnum* aspect	2. S. MACROPHYLLUM
2. Hyaline cells with evident spiral thickenings, pores various; plants usually with typical *Sphagnum* aspect	3
3. Green cells entirely included by hyaline cells, not exposed on either surface of leaf; plants sometimes with purple-red tinge	4
3. Green cells exposed on one or both surfaces of leaf; plants with or without purple-red tinge	5
4. Cortical cells of stems with spiral thickenings	3. S. MAGELLANICUM
4. Cortical cells of stems lacking spiral thickenings	11. S. COMPACTUM
5. Cortical cells of branches with spiral thickenings (thickenings	

sometimes faint and difficult to see); resorption furrow present
in leaf margin 6
5. Cortical cells of branches lacking spiral thickenings; resorption
furrow present or lacking 10
   6. Green cells rectangular to elliptical, exposed about equally on
both surfaces of leaf .... 4. S. PERICHAETIALE
   6. Green cells triangular (sometimes trapezoidal), exposed most
broadly on inner surface of leaf 7
7. Cortical cells of branches with their lower end-walls obconical
and protruding into the next cell below .... 5. S. PORTORICENSE
7. Cortical cells of branches with flat end-walls that do not extend
into the cells below 8
   8. Green cells equilateral-triangular .... 6. S. IMBRICATUM
   8. Green cells isosceles-triangular, rarely trapezoidal 9
9. Dorsal surface of hyaline cells in leaf base mostly with 1–5
pores, the pores elliptic, mostly 24–38 μm long; walls of
hyaline cells adjacent to green cells smooth .... 7. S. PALUSTRE
9. Dorsal surface of hyaline cells in leaf base mostly with 7–16
pores, the pores circular, mostly 10–18 μm diameter; walls of
hyaline cells adjacent to green cells usually ridged (best seen
in longitudinal sections of the leaves) .... 8. S. HENRYENSE
   10. Cortical cells of branches essentially all alike, most with a
pore at the upper end; resorption furrow present .... 10. S. STRICTUM
   10. Cortical cells of branches of two kinds, some with a pore at
the upper end and some without a pore; resorption furrow
mostly absent 11
11. Green cells exposed most broadly on inner surface of leaf, triangular to trapezoidal; plants often tinged pink or red 12
11. Green cells exposed most broadly on outer surface of leaf, triangular to trapezoidal, or rectangular and with about equal
exposure on both surfaces; plants green or brownish, lacking
pink-red tinge 14
   12. Margins of branch leaves ± serrulate toward leaf tip ...... 13. S. MOLLE
   12. Margins of branch leaves entire (leaf apex may be toothed) ... 13
13. Hyaline cells of stem leaves mostly lacking pores on outer surface; outer surface of hyaline cells in apical part of leaf with
many small ± circular pores .... 12. S. BARTLETTIANUM
13. Hyaline cells of stem leaves with pores on outer surface; outer
surface of hyaline cells in apical part of leaf with a few large
elliptic pores .... 14. S. TENERUM
   14. Branch leaves wavy, tips ± twisted and recurved when dry;
cortical cells of stem not or poorly differentiated from in-
ner cells; green cells triangular .... 15. S. RECURVUM
   14. Branch leaves not particularly wavy, twisted, or recurved
when dry; cortical cells of stem usually enlarged,
hyaline, and well-differentiated from inner cells (but
sometimes small and not greatly differentiated in *S.
cuspidatum*); green cells triangular to trapezoidal or
rectangular 15
15. Hyaline cells typically with numerous small pores along their
sides dorsally, giving a beadlike appearance, at least in upper

part of leaf, the pores sometimes few; branch leaves concave
when dry   9. S. LESCURII
15. Hyaline cells with the pores fewer and not beadlike in appearance or arrangement; branch leaves not strongly concave when dry   16
    16. Branch leaves short-ovate to oblong-rectangular, serrulate
on margins, apex of at least some broad and blunt   16. S. FITZGERALDII
    16. Branch leaves lanceolate to linear-lanceolate, entire or ser-
rulate on margins, apex acute   17
17. Leaf margins entire below the toothed tip; leaves lanceolate to
long-lanceolate   17. S. CUSPIDATUM
17. Leaf margins serrulate; leaves linear-lanceolate, tapering to a
long, slender tip   18. S. TRINITENSE

---

1. **Sphagnum cyclophyllum**
Sull. & Lesq. ex Sull.
Figure 3, A–C

Plants rather small for the genus, mostly brownish green or paler but sometimes with pink tinge, in soft cushions or floating mats; stems turgid, mostly 3–4 cm long, mostly unbranched, occasion‐ally with one or several branches but the branches not in clusters; leaves all alike, very concave, not much longer than wide; green cells trapezoidal to rectangular or triangular, with broadest ex‐posure on inner surface or about equal exposure on both surfaces; hyaline cells with numerous small pores on outer surface along edges of cells; resorption furrow absent.

> **Habitat:** Often aquatic in wet pine savannahs, flatwoods, ditches, ponds; sometimes abundant in shallow, wet depres‐sions in wet, grassy savannahs.
> **Range:** Atlantic and Gulf coastal plains in North America; Brazil.
> **Gulf Coast Distribution:** Florida to eastern Texas.

Plants of this species are distinctly unlike most *Sphagnum* in appearance because they lack the one-sided clusters of branches along the stems typically present in most other species of peat mosses. Nonetheless, the leaves are of the regular *Sphagnum* type. When seen floating in water, the turgid stems of *S. cyclophyllum* sometimes bear a surprising resemblance to worms. The pores on the outer surface of the hyaline cells of the leaves are very similar to those found in *S. lescurii*, but are generally smaller. Little-branched forms of the latter species may resemble *S. cyclophyllum*.

2. **Sphagnum macrophyllum**
Bernh. ex Brid.
Figure 3, D–F

Rather coarse plants, generally resembling small, tufted grasses or sedges and lacking the appearance of typical *Sphagnum* species, light to dark green, brownish, or even blackish, often appearing bristly when dry due to the pointed leaves projecting in all direc‐tions; commonly floating; hyaline cells of leaves lacking spiral thickenings, bearing a median row—or two rows—of small pores.

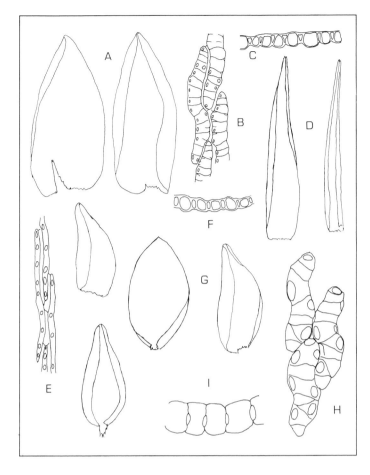

Fig. 3. *A–C:* **Sphagnum cyclophyllum.** *A,* branch leaves (× 15). *B,* cells of branch leaf, outer surface (× 220). *C,* cells of branch leaf, cross section (× 220). *D–F:* **Sphagnum macrophyllum.** *D,* branch leaves (× 11). *E,* cells of branch leaf, outer surface (× 220). *F,* cells of branch leaf, cross section (× 220). *G–I:* **Sphagnum magellanicum.** *G,* branch leaves (× 15). *H,* cells of branch leaf, outer surface (× 220). *I,* cells of branch leaves, cross section (× 220).

Habitat: Pools and wet depressions in pine savannahs and flatwoods; streams and swamps; floating or stranded.

Range: Eastern North America, mostly coastal plain.

Gulf Coast Distribution: Florida to eastern Texas.

Sphagnum macrophyllum is our most easily recognized species of the genus, once it has been learned. It resembles no other *Sphagnum* in our area. Its peculiar bristly (when dry) habit makes it look like a small vascular plant, such as a grass or a sedge, and thus it may be passed over in the field and not recognized as a moss. Once learned, however, it is easily known in the field. Under the microscope the absence of spiral thickenings in the hyaline cells of the leaf and the distinctive linear arrangement of the pores in one or two series are characteristic. This species is much more common in the eastern part of the Gulf Coast than to the west. It is rare in Louisiana and Texas.

Plants of this species whose leaves bear two rows of pores may be recognized as *S. macrophyllum* var. *floridanum* Aust. It is apparently not known along the Gulf Coast west of Florida.

3. Sphagnum magellanicum
Brid.
Figure 3, G–I

Robust, usually reddish or purplish plants (or tinged with color only at the stem tips), to 15 cm or more tall; branches stout, turgid, leaves somewhat loosely inserted; green cells in cross section entirely included within the hyaline cells and not exposed on either surface; hyaline cells with a few large pores on the outer surface; cortical cells of the branches with spiral thickenings; resorption furrow present on leaf margins.

> **Habitat:** Around springs, stream banks, pond edges, swamps.
> **Range:** North and South America; Europe; Asia.
> **Gulf Coast Distribution:** Florida to eastern Texas.

These robust, usually colored plants are easy to identify, especially when the entirely included green cells are seen in cross sections of the leaf. *Sphagnum compactum* shares this feature but differs in obvious ways; see comments under that species. Our only other coarse, sometimes robust *Sphagnum* that may have red-purple coloration is *S. perichaetiale*, which has short, often indistinct branches and a compact growth habit. As is the usual case for Gulf Coast *Sphagnum* species, *S. magellanicum* is more common along the Gulf to the east than it is to the west.

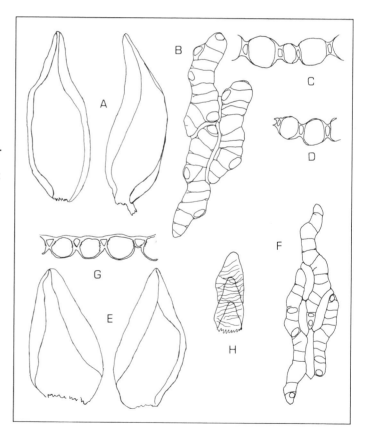

Fig. 4. *A–D:* **Sphagnum perichaetiale.** *A*, branch leaves (× 15). *B*, cells of branch leaf, outer surface (× 220). *C–D:* cells of branch leaves, cross section. *D*, showing resorption furrow (× 220). *E–H:* **Sphagnum portoricense.** *E*, branch leaves (× 15). *F*, cells of branch leaf, outer surface (× 220). *G*, cells of branch leaf, cross section (× 220). *H*, cortical cells of branch (× 55).

Plants small to fairly robust, rather coarse, usually in compact clumps, sometimes in loose mats, green or with brown to red-purple tinge, sometimes dark brown; stems mostly not much elongated; branches crowded, short, and often individually indistinct due to compact growth habit; green cells ± rectangular or lenticular, exposed about equally on both surfaces of leaf; hyaline cells with several small to medium pores on outer surface; cortical cells with spiral thickenings; resorption furrow present.

> **Habitat:** Pools and depressions in wet pine savannahs and flatwoods; shrubby bogs; edges of swamps, ponds, ditches.
>
> **Range:** Coastal plain of eastern North America from New Jersey south; widespread in other parts of the world including the American tropics, Asia, and elsewhere.
>
> **Gulf Coast Distribution:** Florida to eastern Texas.

The rather coarse plants growing in usually compact clumps and the indistinct branches make this moss fairly easy to recognize in the field. Under the microscope the equally exposed green cells of the leaves (as seen in cross section) and the spiral thickenings (sometimes faint) of the cortical cells of stems and branches identify *S. perichaetiale*. It is a common plant in wet pine forests in the Florida Panhandle and along the Gulf to southeastern Louisiana, but is rare further west.

Mostly rather coarse, robust plants with brownish tinge, growing in small clumps, rarely abundant; branches stout, tumid, often curved downwards, elongate; green cells equilateral-triangular in section, exposed on inner surface of leaf; hyaline cells with a few large pores on outer surface, often with small ridges on their walls adjacent to the green cells; cortical cells of branches with spiral thickenings, the end wall of each bulging downward into the next cell below; resorption furrow present.

> **Habitat:** Swamps, springs, edges of ponds and streams, ditches, wet places in woods.
>
> **Range:** Coastal plain of eastern North America from Long Island toward the south; Puerto Rico, Guadeloupe; Mexico; northern South America.
>
> **Gulf Coast Distribution:** Florida to southeastern Louisiana (St. Tammany and Washington parishes).

The robust stature, brownish color, and smooth, tumid, downwardly curved branches are fairly diagnostic field characteristics for *S. portoricense*. Under the microscope the funnel-shaped end-walls of the cortical cells of the branches, protruding down into the cells below, are also characteristic. This species is rather uncommon and only rarely abundant.

4. **Sphagnum perichaetiale** Hampe
Figure 4, A–D

Sphagnum erythrocalyx Hampe (in the sense of Breen, 1963).

5. **Sphagnum portoricense** Hampe
Figure 4, E–H

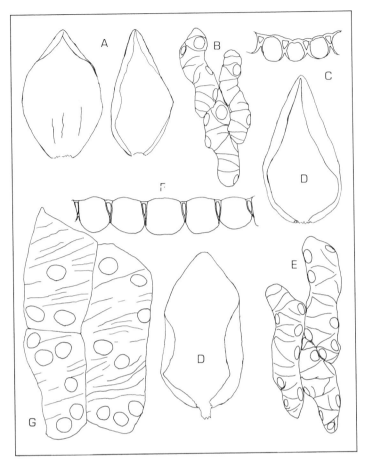

Fig. 5. *A–C:* **Sphagnum imbricatum.** *A,* branch leaves (× 15). *B,* cells of branch leaf, outer surface (× 220). *C,* cells of branch leaf, cross section (× 220). *D–G:* **Sphagnum palustre.** *D,* branch leaves (× 15). *E,* cells of branch leaf, outer surface (× 220). *F,* cells of branch leaf, cross section (× 220). *G,* cortical cells of stem (× 220).

6. Sphagnum imbricatum
Hornsch. ex Russ.
Figure 5, A–C

Rather robust, coarse plants, often in dense clumps or mats, greenish to tinged with brown; branches usually elongated, stout, often loosely foliated; green cells equilateral-triangular in section, exposed on inner surface of leaf; hyaline cells with several elliptical pores on the outer surface, usually with strong ridges on their walls adjacent to the green cells, especially towards the base of the leaf, the ridges appearing as papillae in the cross section and as comb-like features on the outer surface of the leaf; cortical cells of branches with spiral thickenings; resorption furrow present in the leaf margin.

> **Habitat:** Savannahs, wet forests, springs, stream banks, pond margins; common and often abundant.
> **Range:** Eastern North America, mainly on the coastal plain but here and there in the interior; Alaska; Cuba; Europe; Asia.
> **Gulf Coast Distribution:** Florida to eastern Texas.

This moss is one of our most common *Sphagnum* species and thus frequently collected; it sometimes covers large areas in low, wet woods. The combination of equilateral-triangular green cells, spiral thickenings in the cortical cells of branches, and green to brownish coloration should distinguish this species from similar forms. The comblike ridges, which are usually present, are also helpful in distinguishing *S. imbricatum*, but they are sometimes absent or very much reduced. *Sphagnum palustre* and *S. henryense* are superficially similar to *S. imbricatum*, but are easily distinguished by the characteristics cited in the key.

Rather robust, coarse, green plants sometimes tinged with brown, in clumps or thin mats; branches well developed, usually loosely foliated; green cells narrowly isosceles-triangular in section (rarely trapezoidal), exposed on the inner side of the leaf; hyaline cells mostly with 1–5 elliptic pores on the dorsal surface, the pores mostly 24–28 μm long; walls of hyaline cells smooth adjacent to green cells (as seen in section); cortical cells of branches with spiral thickenings; resorption furrow present.

7. Sphagnum palustre L.
Figure 5, D–G

> **Habitat:** Primarily a forest moss; along streams and in seepy areas, around springs and wet depressions; swamps.
> **Range:** Eastern North America; Mexico; Europe; Asia.
> **Gulf Coast Distribution:** Florida to eastern Texas.

Although widely distributed in our area, *S. palustre* is not really a common moss, much less so, for example, than *S. imbricatum* or *S. lescurii*. It is similar to *S. henryense* but can usually be distinguished by the characters cited in the key. It differs from *S. imbricatum*, among other ways, in its isosceles-triangular green cells and in having the walls of the hyaline cells smooth on their sides adjacent to the green cells.

Similar in general appearance to *S. palustre* and perhaps not always distinguishable; differing most obviously from that species by the more numerous, smaller, circular pores in the hyaline cells on the dorsal surface of the leaf, especially at the leaf base, and in the reticulate-ridged lateral walls of the hyaline cells adjacent to the green cells. The ridges are most easily seen in longitudinal sections of the leaves; they appear as small projections of the cell walls in cross section.

8. Sphagnum henryense Warnst.
Figure 6, A–F

> **Habitat:** Wet forests; around springs and seeps; along streams.
> **Range:** Eastern United States, northwestern North America; Cuba; Japan.
> **Gulf Coast Distribution:** Florida, Alabama, Louisiana, eastern Texas.

Sphagnum henryense is apparently not a common species in our area and not yet reported from Mississippi. See Andrus (1974) for further discussion on the distinctions between *S. henryense* and *S. palustre*.

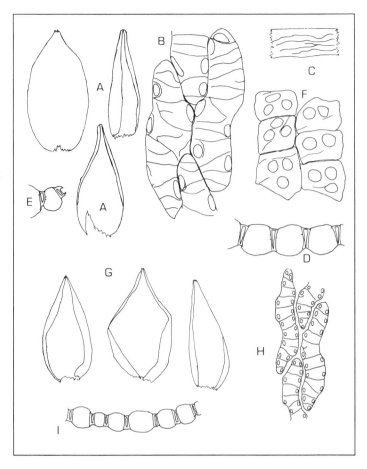

Fig. 6. *A–F:* **Sphagnum henryense.** *A,* branch leaves (× 15). *B,* cells of branch leaf, outer surface (× 220). *C,* portion of hyaline cell of branch leaf, horizontal section, showing linear thickenings (× 231). *D–E,* cells of branch leaves, cross section. *E,* showing resorption furrow (× 220). *F,* cortical cells of stem (× 220). *G–I:* **Sphagnum lescurii.** *G,* branch leaves (× 15). *H,* cells of branch leaf, outer surface (× 220). *I,* cells of branch leaf, cross section (× 220).

9. **Sphagnum lescurii** Sull. in Gray
Figure 6, G–I

Plants quite variable in size and appearance, mostly slender but sometimes robust, green to yellowish green or paler, sometimes brownish, often with a metallic sheen, sometimes nearly isophyllous and with few branches, in thin mats or dense clumps; branches mostly short and pointed, sometimes indistinct; green cells small, trapezoidal to almost rectangular, exposed most broadly on outer surface or with about equal exposure on both surfaces; hyaline cells of branch leaves typically with numerous small pores along their edges resembling chains of beads, at least toward the leaf tip, the pores sometimes few; cortical cells of branches lacking spiral thickenings; resorption furrow lacking.

> **Habitat:** Ditch banks, wet forests, pond margins, ravines.
> **Range:** Eastern North America.
> **Gulf Coast Distribution:** Florida to eastern Texas.

This species is very common and often abundant in some parts of its Gulf Coast range, particularly in Louisiana. Although quite variable in many qualities, it can usually be easily identified by the small pores in beadlike series on the outer surface of the hyaline cells of the branch leaves. These pores are present in typical form in at least some leaves of virtually every specimen, although sometimes best developed only toward the tips of the leaves. Some specimens approach *S. cyclophyllum* in form in being almost isophyllous and in having few branches. *Sphagnum lescurii* is probably the most frequently collected peat moss in the central part of the Gulf Coast. It was included for many years in the concept of *S. subsecundum* but is generally considered now to be distinct. True *S. subsecundum* has a more northerly distribution in the United States and apparently does not occur along the Gulf Coast. *Sphagnum lescurii* is sometimes treated as a taxonomic variety of *S. subsecundum*.

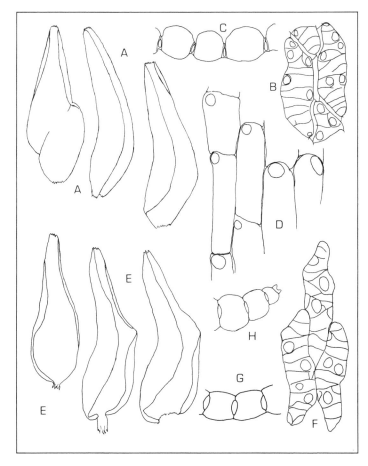

Fig. 7. *A–D*: **Sphagnum strictum.** *A*, branch leaves (× 15). *B*, cells of branch leaf, outer surface (× 220). *C*, cells of branch leaf, cross section (× 220). *D*, cortical cells of branch (× 220). *E–H*: **Sphagnum compactum.** *E*, branch leaves (× 15). *F*, cells of branch leaf, outer surface (× 231). *G*, cells of branch leaf, cross section (× 231). *H*, margin of branch leaf, cross section, showing resorption furrow (× 231).

Small, mostly very compact plants with green to yellowish green coloration; internodes often short and branches thus crowded on stems; branches very short; green cells rather narrowly isosceles-triangular, exposed on the outer surface of the leaf; hyaline cells with few to many small, ringed pores along their edges; cortical

10. **Sphagnum strictum** Sull.
Figure 7, A–D

cells of branches without spiral thickenings, most with a pore at the upper end; resorption furrow present.

> **Habitat:** Flatwoods, ditches, swamps, wet forests.
> **Range:** Eastern North America; Mexico; parts of the American tropics; Europe.
> **Gulf Coast Distribution:** Florida, Alabama, Louisiana; not yet reported from Mississippi but almost surely present there.

This species is fairly common in Florida, where it is frequently found with sporophytes. Its small size makes it inconspicuous. The essentially uniform cortical cells of the branches, most of which bear a pore at the upper end, are diagnostic for this species. The small size and compact growth habit are also helpful in identification, although occasional colonies have plants which are larger and less compact than in the common condition. See comments under *S. compactum*, a similar species.

11. **Sphagnum compactum**
DC ex Lam. & DC
Figure 7, E–H

Plants rather small and compact, pale yellowish green, sometimes tinged with brown; branches short, crowded on stem; green cells entirely included by hyaline cells, not exposed on either surface of the leaf; outer surface of hyaline cells with a few large pores; cortical cells of branches lacking spiral thickenings, most with a prominent pore at the upper end; resorption furrow present on leaf margins.

> **Habitat:** Wet areas in forests; wet rock ledges.
> **Range:** Widespread in the Northern Hemisphere
> **Gulf Coast Distribution:** Attributed to Louisiana by Crum and Anderson (1981), but I have not seen a specimen.

Sphagnum compactum is generally similar in appearance to *S. strictum*, a near relative, but can easily be distinguished by having the green cells entirely included by the hyaline cells, as seen in cross section. Our only other *Sphagnum* with this feature is *S. magellanicum*. Plants of the latter species, however, are robust, red-tinged, and have the cortical cells of stems and branches with spiral thickenings, among other differences.

12. **Sphagnum bartlettianum**
Warnst.
Figure 8, A–C

Plants slender, rather delicate, elongate, in dense clumps, mostly tinged with pink-purple in upper parts; branches slender, rather short, curved downward; leaves small, loosely inserted on branches; green cells triangular to trapezoidal in section, mostly $\frac{1}{2}$ or more height of hyaline cells, with broadest exposure on inner surface; hyaline cells with several to many small, strongly ringed pores on outer surface, especially in upper part of leaf; cortical cells of branches without spiral thickenings; resorption furrow lacking.

> **Habitat:** Shrubby bogs; swamps.
> **Range:** Southeastern United States.

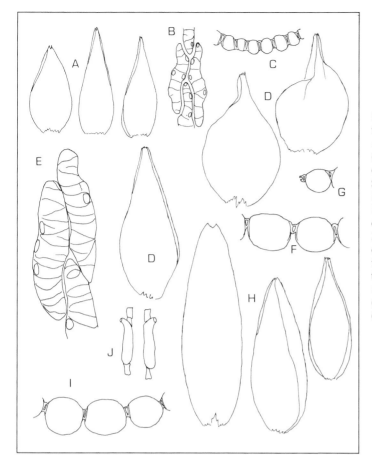

Fig. 8. *A–C:* **Sphagnum bartlettianum.** *A,* branch leaves (× 26). *B,* cells of branch leaf, outer surface (× 220). *C,* cells of branch leaf, cross section (× 220). *D–G:* **Sphagnum molle.** *D,* branch leaves (× 15). *E,* cells of branch leaf, outer surface (X 220). *F,* cells of branch leaf, cross section (× 220). *G,* margin of branch leaf, cross section, showing resorption furrow (× 220). *H–J:* **Sphagnum tenerum.** *H,* branch leaves, (× 26). *I,* cells of branch leaf, cross section (× 220). *J,* cortical cells of branch (× 55).

Gulf Coast Distribution: Known from two stations in the Florida Panhandle, in Liberty and Walton counties.

It is possible that this species will turn up eventually in other places in western Florida as well as further west along the Gulf Coast. It was originally reported from Florida under the name *S. warnstorfii* Russ., a species with a much more northerly range (Andrus 1979).

Plants in loose or compact clumps, pale green to yellowish or nearly white, often tinged with pink or red; branches ascending, sometimes indistinct; green cells triangular to trapezoidal, exposed most broadly on inner side; hyaline cells with a few elliptical or rounded pores along their sides and sometimes a few rounded pores in the middle, pores smaller and more numerous towards leaf apex; at least some branch leaves toothed on the margins toward the leaf apex; cortical cells of branches without spiral thickenings; resorption furrow often present.

13. **Sphagnum molle** Sull.
Figure 8, D–G

Sphagnum tabulare Sull.

Habitat: Wet depressions in savannahs; swamps, ditches.

Range: Eastern North America, mainly near the coast; western Europe.

Gulf Coast Distribution: Florida to eastern Texas.

The toothed margin is not present on every leaf but is evident when found. The usually compact habit with the branches ascending, and the pink or reddish tinge, are helpful in identification.

14. **Sphagnum tenerum** Sull. & Lesq. ex Sull.
Figures 8, H–J; 9, A

Plants delicate to rather robust, sometimes very compact, mostly with elongate stems, pale yellowish green or tinged with pink-red, in dense or loose clumps; branches distinct, mostly short, sometimes elongated and pointed downwards; green cells tiny, exposed most broadly on inner surface, isosceles-triangular to trapezoidal, sometimes rectangular; hyaline cells with few, mostly elliptical pores along their edges on the outer surface; leaf margins entire; cortical cells without spiral thickenings, pores with usually prominent, protruding necks; resorption furrow absent.

Habitat: Wet depressions in savannahs, pond margins, ditches; wet humus in swamps.

Range: Eastern North America.

Gulf Coast Distribution: Florida, Alabama. Not yet reported from Mississippi but probably occurring there.

This species has been collected a number of times in northern Florida and in the Panhandle along the Gulf. The usually prominent, protruding neck of the cortical cell pores is helpful in identifying *S. tenerum*.

15. **Sphagnum recurvum** P.-Beauv.
Figure 9, B–E

Plants robust, graceful, green to yellowish, in large or small clumps; branches elongated, often appearing furry when dry because of spreading leaf tips; green cells triangular in section, exposed on the outer surface; hyaline cells with few pores on the outer surface; cortical cells without spiral thickenings; resorption furrow absent; cortical cells of stem not very well differentiated from inner cells as seen in section.

Habitat: Bogs, around springs and ponds, wet forests.

Range: North America south to California, Missouri, and the Gulf Coast; Mexico; Costa Rica; Panama; South America; Europe; Asia.

Gulf Coast Distribution: Florida to eastern Texas; more common in the eastern part of the Gulf Coast than in the west; rare in Louisiana and Texas.

Sphagnum recurvum, at least as it occurs along the Gulf Coast, is not likely to be confused with any other species of *Sphagnum*. Its robust stature, elongated branches with spreading leaf tips, triangular green cells, and poorly differentiated cortical cells on the stems

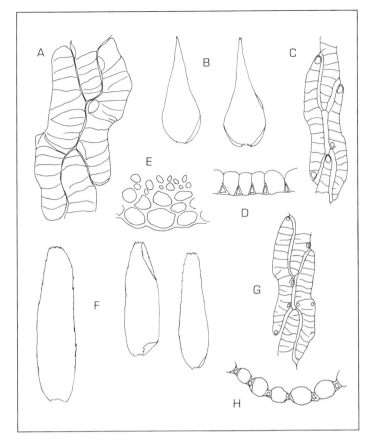

Fig. 9. *A*, **Sphagnum tenerum**, cells of branch leaf, outer surface (× 220). *B–E*: **Sphagnum recurvum.** *B*, branch leaves (× 15). *C*, cells of branch leaf, outer surface (× 220). *D*, cells of branch leaf, cross section (× 220). *E*, portion of cross section of stem (× 220). *F–H*: **Sphagnum fitzgeraldii.** *F*, branch leaf, (× 15). *G*, cells of branch leaves, outer surface (× 220). *H*, cells of branch leaf, cross section (× 220).

combine to make it easy to identify. It is also easy to recognize in the field.

Plants slender, rather delicate, pale green to whitish, in thin mats or small clumps, often aquatic; branches slender, short, often indistinct; branch leaves serrulate on their margins, often broad and very blunt at the toothed apex, ovate to oblong or nearly rectangular in outline; green cells tiny, trapezoidal, exposed most broadly on outer surface; hyaline cells with few pores; cortical cells of branches without spiral thickenings; resorption furrow absent.

16. Sphagnum fitzgeraldii
Ren. ex Lesq. & James
Figure 9, F–H

> **Habitat:** Pond edges, wet depressions in pine forests, shrub bogs.
> **Range:** North America; mostly on the coastal plain from North Carolina south.
> **Gulf Coast Distribution:** Florida and Alabama.

Sphagnum fitzgeraldii is not a common moss and this, together with its small size, results in only infrequent collections of it. It is more delicate in appearance than *S. lescurii* and lacks the numerous small pores of that species; also its leaf tips are broad and blunt and the leaves are not as pointed. This species is best recognized by the oblong-rectangular shape of the branch leaves; not all of the branch leaves may be oblong-rectangular, but some of them will have the characteristic shape. In most collections the stems are elongate, but in some the plants are very compact.

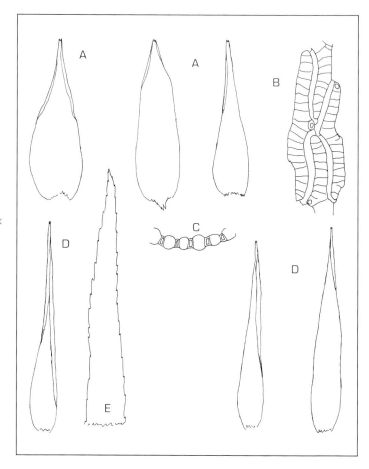

Fig. 10. *A–C:* **Sphagnum cuspidatum.** *A,* branch leaves (× 26). *B,* cells of branch leaf, outer surface (× 220). *C,* cells of branch leaf, cross section (×220). *D–E:* **Sphagnum trinitense.** *D,* branch leaves (× 15). *E,* tip of branch leaf (× 55).

17. **Sphagnum cuspidatum**
Ehrh. ex Hoffm.
Figure 10, A–C

Plants pale green to whitish or yellowish, usually glossy, elongate, slender, soft and lax when floating, firmer and more compact when stranded; branches usually slender, tapering; branch leaves tapering to an acute apex, somewhat undulate and not strongly concave when dry, margins entire; green cells triangular to trapezoidal or rectangular, mostly exposed most broadly on outer surface but sometimes with ± equal exposure, about same height as hyaline cells as seen in section or a little shorter; hyaline cells with few, rather small ringed pores on outer surface or pores essentially lacking; cortical cells of branches without spiral thickenings; resorption furrow lacking.

Habitat: Ditches, ponds, depressions in pine forests; mostly aquatic and floating but often stranded by falling water levels.

Range: Eastern North America; Europe; Asia.

Gulf Coast Distribution: Florida to Mississippi.

This moss is not common along the Gulf Coast, or at least not often collected. Its generally aquatic habitat, soft, elongate stems, and green cells only a little shorter than adjacent hyaline cells combine to make it recognizable. Under the dissecting microscope the plants have a characteristic glossy aspect. Stranded plants are more compact and may have a quite different aspect than floating ones. Smith (1978) noted that the flaccid habit of this plant "has aptly been likened to a drowned kitten."

Similar in many respects to *S. cuspidatum* but differing in that the branch leaves have much longer and narrower tips, distinctly serrulate on their margins.

Habitat: Swamps, ditches, ponds, depressions in pine forests; mostly aquatic and floating.

Range: Eastern North America, mainly along the coast; Bermuda; Puerto Rico; South America; Europe.

Gulf Coast Distribution: Florida to southeastern Louisiana.

As in *S. cuspidatum*, stranded plants of *S. trinitense* are more compact and firmer than floating ones; the branch leaves of stranded plants are shorter and broader than those of floating plants.

18. **Sphagnum trinitense** C.M.
Figure 10, D–E

Sphagnum cuspidatum Ehrh. ex Hoffm. var. *serrulatum* (Schleiph.) Schleiph.

SPHAGNUM PLUMULOSUM Röll (*S. subnitens* Russ. & Warnst. ex Warnst.) was reported from Harrison County, Mississippi, by Wilkes (1963). However, I have seen this specimen, and it is actually *S. molle*.

SPHAGNUM TORREYANUM Sull. [*S. cuspidatum* var. *torreyanum* (Sull.) Braithw.] has been reported from Louisiana, but the specimen upon which the report was based is actually *S. cuspidatum* (L. E. Anderson 1980: personal communication).

FISSIDENTACEAE Schimp.

Tiny to rather large plants, yellowish to green or darker, sometimes glossy, gregarious or in loose to compact sods; stems mostly erect, simple or sparingly branched; leaves oblong-lanceolate, sometimes bordered in part or all around by elongate cells or by cells of a different color; leaves in two rows on opposite sides of the stems so that the leafy stems are flattened; leaves each consisting of a dorsal lamina, an apical lamina, and two vaginant laminae, the last clasping the stem (Fig. 11); cells mostly ± isodiametric, smooth to papillose; costa single, rarely lacking. Sporophytes terminal or lateral; setae relatively short, often curved; capsules exserted, erect or inclined; operculum rostrate; peristome single, the teeth mostly divided ⅓-⅔ their length; calyptra cucullate.

All of our species are treated under the single genus, *Fissidens*.

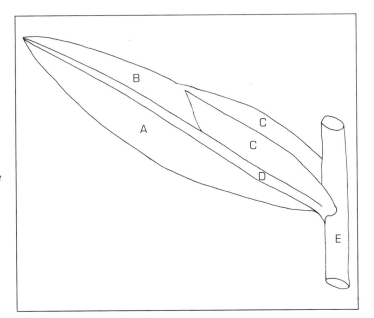

Fig. 11. Diagram of a single leaf of a **Fissidens** plant, attached to a segment of the stem. *A*, dorsal lamina. *B*, apical lamina. *C*, vaginant laminae. *D*, costa. *E*, stem segment. The vaginant laminae clasp the stem between their bases.

FISSIDENS Hedw.

The name *Fissidens* means "split tooth," referring to the usually divided peristome teeth. These plants are common and often abundant throughout most of the Gulf Coast, although only rarely conspicuous because most of them are rather small. Some of our species grow on soil, some on rocks, and some on bark. Two of our species are aquatic, growing in flowing or standing water and attached to rocks, roots, logs, and tree bases. The genus *Fissidens* is one of the easiest to recognize and remember because of the form of the leaves and their arrangement on the stem. The lower portion of each leaf (Fig. 11) is split to the costa on its upper side to half or more the length of the leaf, with the split portion clasping the stem at the point of attachment, and often the base of the leaf above. This feature alone is distinctive because it occurs in no other Gulf Coast moss; it can usually be observed in the field with a hand lens. The distichous leaf arrangement is also distinctive among our mosses. Each leafy stem of *Fissidens* is flat (at least when moist), like the frond of a fern, and indeed the plants of many species of the genus do resemble minute ferns. In some of the species the fernlike aspect is evident even when the plants are dry, but in others the plants become variously contorted when dry and the fernlike appearance is distorted or lost altogether.

Common habitats for *Fissidens* include stream banks, shady lawns, damp rocks, walls and ledges, tree bases, and bare soil in general. In favorable habitats extensive colonies as large as a square meter or more may develop.

As is the case for various other genera of Gulf Coast mosses, our species of *Fissidens* include representatives of the eastern North American bryoflora as well as representatives from the tropics. Some of the species are more or less ubiquitous, occurring widely around the world.

Some *Fissidens* species have tough, persistent plants that endure for more than one season, but other species flourish during the winter-spring season and then disappear during the warm summer months. Survival and reestablishment of the latter are probably due to a combination of regeneration from rhizoids and buried stems and leaves, and growth from spores.

Most species of *Fissidens* produce spores in abundance. As in other mosses, dispersal of spores from the capsule is moderated by the peristome. Some *Fissidens*, members of the *F. bryoides* group, for example, make fine subjects for observing the hygroscopic movements of the peristome. If one places a clump of fruiting *Fissidens* plants under a dissecting microscope and then gently blows on the capsules, the peristome teeth will respond by gracefully and quickly moving in and out in response to the alternation of moist breath with the drier ambient air.

Some *Fissidens* species have tiny plants, while the plants of others grow to be quite large. Plants of *F. hyalinus* and *F. neonii*, for example, have minute stems just a few millimeters tall and may be virtually invisible to a standing observer. Others, such as *F. polypodioides* and *F. asplenioides*, are much more robust, up to several centimeters tall and easily visible from a standing position.

Fissidens plants commonly grow in dense colonies, with the flattened fronds all parallel to one another so that, when viewed from an appropriate aspect, the fronds are seen to overlap like shingles. Extensive areas of steep, otherwise bare soil banks may be covered in this manner, with the *Fissidens* plants forming a protective cover that retards erosion. Conical, meter-high termite mounds in the forests of the Amazon Basin may be covered by such green shingling, often composed of two or more species of *Fissidens*. Presumably the shingling contributes toward stability of the mound surface by protection from the heavy tropical rains. Crayfish "castles" in central Gulf swamplands often display a similar covering of *Fissidens* plants, although usually not as well developed as on the termite mounds of Amazonia.

Special Study Methods for *Fissidens*

Fissidens plants are easy to study because the plants are mostly small. They can be lifted from the substrate with forceps and placed in water on a microscope slide very conveniently. Pick out several plants to place on the slide. *Fissidens* species often grow in mixture, so look over the whole specimen to see if more than one type of plant is present.

In order to determine whether or not the leaf cells are flat and smooth, bulging, or papillose, you must see the leaf either in cross section or at a point where it is folded (where you can see the cells in profile along the fold). In the species with small, pluripapillose cells, the papillae can usually be seen in profile on the marginal cells of the leaf. If you do not want to cut cross sections of the leaves to see the nature of the cells, try dropping several dry *Fissidens* plants into water on a slide and then quickly dropping the cover slip on top of the plants before they have a chance to soak up and spread out. In this way some of the leaves will remain folded and you can check for papillae along the folds.

1. Leaves ecostate; plants minute, 2–3 mm tall	1. F. HYALINUS
1. Leaves conspicuously costate; plants mostly taller than 2–3 mm	2
2. Leaf cells densely pluripapillose, obscure; border of elongate cells often present on vaginant laminae of upper and perichaetial leaves	3
2. Leaf cells smooth, bulging, or unipapillose, mostly distinct; border of elongate cells present or lacking	4
3. Costa ending well below leaf apex	2. F. GARBERI
3. Costa percurrent	3. F. RAVENELII
4. Leaves, at least the upper ones, entirely or partly bordered with elongate cells	5
4. Leaves lacking border of elongate cells	10
5. Leaf cells each with a single, small papilla; border mostly confined to vaginant laminae; habitat mainly swamps	6
5. Leaf cells smooth and flat to bulging, not papillose; border confined to vaginant laminae or more extensive; habitat various	7
6. Leaves ± apiculate; border on vaginant laminae narrow, 2–3 cells wide; mostly on soil	4. F. KOCHII
6. Leaves not particularly apiculate; border on vaginant laminae up to 5 cells wide; mostly on tree bases	5. F. REESEI
7. Leaves bluntly rounded at tips; costa ending below leaf apex; border of elongate cells weak, on vaginant laminae of upper leaves only	6. F. OBTUSIFOLIUS
7. Leaves pointed at apex; costa percurrent-excurrent; border of elongate cells complete or less extensive, sometimes only on vaginant laminae	8
8. Plants tiny, 1–2 mm tall; border of elongate cells weak	7. F. NEONII
8. Plants more than 2 mm tall; border strong, often ± complete	9

9. Cells of vaginant laminae conspicuously larger than those of apical and dorsal laminae; border essentially complete — 9. F. KEGELIANUS

9. Cells of vaginant laminae not obviously different from those of
apical and dorsal laminae; border often incomplete — 8. F. BRYOIDES

    10. Plants very small, 2–3 mm tall; leaves bluntly rounded at
tips; costa usually ending well below the leaf apex — 6. F. OBTUSIFOLIUS

    10. Plants mostly larger; leaves pointed; costa mostly extending
to leaf apex — 11

11. Plants aquatic, soft, attached to various substrates in water (or
exposed by falling water level) — 12

11. Plants terrestrial (although often occurring in wet sites); texture
firm — 13

    12. Sporophytes axillary in lower leaves; peristome teeth
truncate — 19. F. FONTANUS

    12. Sporophytes terminal on major branches; peristome teeth
long and slender — 20. F. MANATEENSIS

13. Costa obscure, covered in upper part of leaf by small, rounded
cells; plants almost exclusively on tree trunks and bases — 12. F. SUBBASILARIS

13. Costa exposed throughout leaf; plants mainly terrestrial — 14

    14. Leaves with pale margins, coarsely and irregularly toothed
at tips (F. adianthoides sometimes has pale leaf margins) — 10. F. CRISTATUS

    14. Leaves without noticeably pale margins, tips toothed or
entire — 15

15. Plants robust, stems to 5 cm tall; leaf tips bluntly rounded to
apiculate; growing on soil or rocks in ravines and along
streams — 16

15. Plants of moderate to small stature, mostly less than 1 cm tall
(to ca 2–3 cm in F. adianthoides); leaves mostly pointed at
tips; habitats various — 17

    16. Leaf cells bulging; costa ending below leaf apex; leaf mar-
gins serrate — 17. F. ASPLENIOIDES

    16. Leaf cells flat; costa ending in or near apiculation; leaf mar-
gins ± entire except at apex — 18. F. POLYPODIOIDES

17. Leaf cells smooth or inconspicuously bulging or unipapillose — 18

17. Leaf cells conspicuously unipapillose or bulging — 21

    18. Leaves contorted when dry; plants dull, coarse; sporo-
phytes lateral — 19

    18. Leaves essentially uncontorted when dry; plants small,
glossy, delicate; sporophytes terminal — 20

19. Leaf apex with stout apiculation, otherwise essentially entire;
cells 7–9 μm diameter — 13. F. TAXIFOLIUS

19. Leaf apex irregularly toothed, lacking a stout apiculation; cells
10–15 μm diameter — 11. F. ADIANTHOIDES

    20. Leaf cells large, to more than 20 μm diameter; margins
from nearly entire to sharply serrate — 15. F. PELLUCIDUS

    20. Leaf cells smaller, to 10 μm diameter; margins finely
serrate-crenate — 16. F. HALLII

21. Leaf apex acute, irregularly toothed; leaves 1.5–4.5 mm long;
cells 10–15 μm diameter — 11. F. ADIANTHOIDES

21. Leaf apex broadly rounded, serrulate to toothed; leaves 1–1.5
mm long; cells 7–9 μm diameter — 14. F. BUSHII

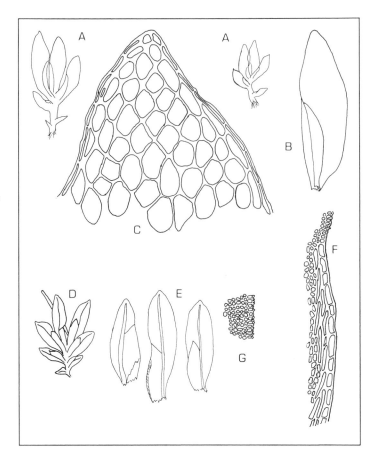

Fig. 12. *A–C:* **Fissidens hyalinus.** *A*, habit (× 11). *B*, leaf, (× 99). *C*, leaf tip (× 231). *D–G:* **Fissidens garberi.** *D*, habit (× 15). *E*, leaves (× 33). *F*, margin of vaginant lamina (× 231). *G*, margin of apical lamina (× 231).

1. **Fissidens hyalinus** Wils. & Hook.
Figure 12, A–C

Minute, delicate plants 1–3 mm tall; leaves ecostate; leaf cells large, to about 70 μm long, the marginal row on each leaf narrow and elongate. Sporophyte terminal.

> **Habitat:** Cool, moist, sheltered places such as deep, forested ravines and damp rock ledges.
> **Range:** Eastern North America (Ohio, Pennsylvania); India; Japan; Mexico; northern South America.
> **Gulf Coast Distribution:** Louisiana (West Feliciana Parish).

Fissidens hyalinus is very rare (or overlooked). On the Gulf Coast it has been found at only one site, in West Feliciana Parish, Louisiana, where it grew among plants of *F. taxifolius* in a deep, wooded ravine. Its striking lack of a costa in the leaves and its very large leaf cells make it easy to identify; no other Gulf Coast *Fissidens* can be confused with it. It is known from only a few other places in North America and elsewhere in the world.

2. **Fissidens garberi** Lesq. & James
Figure 12, D–G

Small, dull plants 2–3 mm tall; leaves about 1 mm long, mostly rather blunt at apex, costa ending several cells below apex; cells pluripapillose, small and obscure; upper leaves of fertile plants with the vaginant laminae bordered by elongate hyaline cells, other leaves mostly unbordered. Sporophyte terminal.

Habitat: Logs, tree bases, rocks in forests, often damp.
Range: Southeastern United States; Missouri, Wisconsin; tropical America.
Gulf Coast Distribution: Florida to Texas.

Fissidens garberi is small and inconspicuous in the field. It is most likely to be confused with the much more common and abundant *F. ravenelii*; however, in *F. ravenelii* the costa goes all the way to the leaf apex rather than ending below it. Also, *F. ravenelii* grows mainly on soil while in our area *F. garberi* grows mainly on tree bases and logs. Both species also occur on rock, sandstone as well as limestone. *Fissidens garberi* and the closely related *F. ravenelii* are sometimes regarded as constituting only a single species, but in our area they appear to be distinct.

Small, dull plants, often in dense colonies; stems 2–4 mm tall; leaves about 1 mm long, the upper ones usually with the vaginant laminae bordered by elongate hyaline cells; cells small, obscure, pluripapillose; costa percurrent. Sporophyte terminal.

3. **Fissidens ravenelii** Sull.
Figure 13, A–B

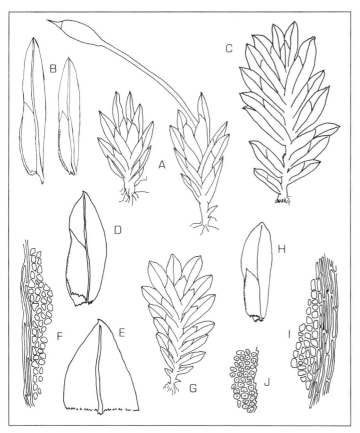

Fig. 13. *A–B*: **Fissidens ravenelii.** *A*, habit (× 15). *B*, leaves (× 33). *C–F*: **Fissidens kochii.** *C*, habit (× 15). *D*, leaf (× 26). *E*, leaf tip (× 55). *F*, margin of vaginant lamina (× 231). *G–J*: **Fissidens reesei.** *G*, habit (× 15). *H*, leaf (× 26). *I*, margin of vaginant lamina (× 231). *J*, margin of apical lamina (× 231).

Habitat: Mostly on soil, also on rock, in open forest.
Range: Southeastern United States; Kansas; Caribbean.
Gulf Coast Distribution: Florida to Texas.

Fissidens ravenelii is a common and weedy moss along the Gulf Coast. It occurs virtually everywhere, usually in rather open sites but also in forests, mostly on soil but also on rocks. Its small stature and dull appearance make it easy to identify in the field with the hand lens. *Fissidens garberi* is also small and dull, and has pluripapillose cells; but its leaves have the costa ending several cells below the leaf tip, and it usually grows on wood. *Fissidens kochii* and *F. reesei* are also small plants that could be confused with *F. ravenelii*; but in both of those species the leaf cells have only a single papilla instead of the several that cells of *F. ravenelii* have. The pluripapillose nature of the cells of *F. ravenelii* is not always easy to see. But by looking at the marginal cells, one can usually see clearly that two or more papillae are present if the plant is *F. ravenelii*.

4. Fissidens kochii Crum & Anderson
Figure 13, C–F

Small, dull, pale green plants, 2–3 mm tall, in thin colonies; leaves about 1 mm long, mostly apiculate, vaginant laminae bordered by 2–3 rows of elongate cells; cells each with a small central papilla. Sporophyte terminal.

Habitat: Clay soil in swamps and low forests.
Range: United States Gulf Coast.
Gulf Coast Distribution: Louisiana.

Along the Gulf Coast this species, because of its unipapillose cells, can only be confused with the similar species *F. reesei*. The latter species, however, differs in having a broader band of elongate cells on the vaginant laminae and in the blunter, nonapiculate leaf apex. *Fissidens kochii* is apparently quite rare; it is known only from a few stations in southern Louisiana, although it is likely that it occurs also further to the east along the coast.

5. Fissidens reesei Crum & Anderson
Figure 13, G–J

Small plants very similar in most respects to the preceding species, *F. kochii*, but differing in the blunter leaf apex and in having 3–5 rows of elongate cells on the margins of the vaginant laminae; the plants are often larger than those of *F. kochii*, up to about 7 mm tall.

Habitat: Muddy tree and stump bases, occasionally on soil, in swamps and along river bottoms.
Range: United States Gulf Coast.
Gulf Coast Distribution: Florida and Louisiana.

Fissidens reesei probably extends into Texas but has not yet been collected there. Although it has not been collected yet in either Mississippi or Alabama, it almost certainly occurs in those states too. Its habit and habitat are such that it is not likely to be

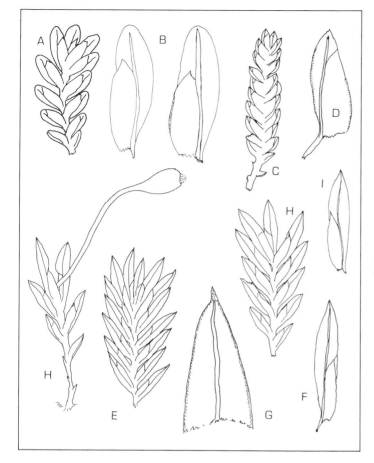

found except by collectors who are willing to seek in the swampy habitats where the species occurs. See comments under *F. kochii.*

Tiny, dull green plants in loose to dense colonies; stems 1–3 mm tall, occasionally to 4 mm; leaves bluntly rounded at tips; costa ending well below leaf apex, often appearing to be forked at its tip; vaginant laminae, especially on upper leaves, often bordered with elongate cells but border completely lacking in many plants; cells smooth. Sporophyte terminal.

6. **Fissidens obtusifolius** Wils.
Figure 14, A–B

> **Habitat:** Damp or wet limestone or other rock in shaded
> places; also on bricks and mortar.
> **Range:** Europe; eastern North America west to the Rockies.
> **Gulf Coast Distribution:** Florida to Texas.

This species is fairly common in limestone areas of Florida and Texas but rather scarce in the intervening area due to lack of suitable substrates. In Louisiana it has been found on old tombs in New Orleans, and at the Locust Grove Cemetery, West Feliciana Parish.

7. Fissidens neonii (Bartr.)
Grout
Figure 14, C–D

Plants tiny, yellowish green, in thin colonies; stems to about 4 mm tall but mostly shorter, only about 0.75 mm wide; leaves about 0.5 mm long, the vaginant laminae occupying ⅔–¾ of the leaf and bordered with elongate cells; dorsal lamina also ± bordered with elongate cells; leaves crowded on stems and almost overlapping, giving the plants a straplike appearance. Sporophyte terminal; peristome teeth undivided.

> **Habitat:** Bare, loesslike soil on steep banks along ravines; often with other mosses including *F. bryoides*, *F. ravenelii*, and *Ditrichum pallidum*.
> **Range:** Louisiana, North Carolina; Mexico.
> **Gulf Coast Distribution:** Southern Louisiana.

This is one of the rarest of Gulf Coast *Fissidens* and the only one in our flora with undivided peristome teeth. It was described as a new species from a specimen collected by Brother Néon near Lafayette, Louisiana, in 1931 and is now known in our area from only a few other localities in Vermilion and West Feliciana parishes. The plants are so small and inconspicuous, and occupy such obscure habitats, that they are difficult to find. The small size, straplike appearance of the stems, relatively large vaginant laminae bordered with elongate cells, and the habitat allow identification. Forms of *F. bryoides* may be similar at first sight but are always larger and have a different habit, not straplike in appearance.

8. Fissidens bryoides Hedw.
Figures 14, E–I; 15, A–B

Fissidens exiguus Sull.;
F. exiguus var. *falcatulus* (Ren. & Card.) Grout; *F. falcatulus* Ren. & Card.; *F. minutulus* Sull.; *F. pusillus* Wils. in Milde; *F. repandus* Wils.; *F. viridulus* (Sw.) Wahlenb. and vars.

Plants green to yellowish green, in dense colonies; stems to about 7 mm tall; leaves 1–2 mm, pointed, broadly to narrowly lanceolate, often a little falcate, bordered entirely or in part with elongate hyaline cells; cells smooth, distinct, mostly about 8–12 μm. Sporophyte terminal.

> **Habitat:** Soil, rock, ledges, tree bases; forests and fields, ravines and banks; often found in yards under trees and around buildings in shady sites.
> **Range:** Widely distributed around the world, mostly in the Northern Hemisphere.
> **Gulf Coast Distribution:** Florida to Texas.

In a sense the name *Fissidens bryoides* is a catchall for a variable assemblage of mosses that is sometimes treated as several species. The important features that permit the various forms to be regarded as a single species are the smooth cells of ± uniform size throughout the leaf and the pointed leaves that are always bordered, at least in part, with elongate cells. Crum and Anderson (1981) give an excellent and compelling review of the reasons for recognizing the several forms encompassed under the name *F. bryoides* as a single entity. Plants treated here as *F. bryoides* vary from quite small, with narrow leaves bordered all around, to larger plants having broad leaves that are bordered only on the vaginant laminae. As a general rule the small plants with completely bordered leaves tend to occur on rock in our area, while the larger ones with various degrees of incompleteness of the border tend to occur mainly on soil.

Plants with broad, spreading, somewhat falcate leaves that are poorly bordered—sometimes only on the vaginant laminae—were originally described as *F. falcatulus*, a taxon that later came to be known as *F. exiguus* var. *falcatulus*. The *falcatulus* form occurs mainly on moist soil and is the most common and abundant expression of *F. bryoides* along the Gulf Coast. Much smaller plants with straight, ascending leaves that usually have a nearly complete border are the *minutulus* form of *F. bryoides*. Such plants occur mainly on rock, along the Gulf Coast, and are quite different in aspect from the *falcatulus* form. The *exiguus* form of *F. bryoides* is similar to the *minutulus* type, but with a much reduced border; and the *viridulus* form is represented by large plants similar to the *falcatulus* expression, but having straight leaves—rather than falcate—and with a nearly complete border. Forms referred to *F. pusillus* and *F. repandus* are other variations on the same general theme.

Fissidens kegelianus is similar at first sight to forms of *F. bryoides* with completely bordered leaves, but is easily distinguished. See comments under *F. kegelianus*.

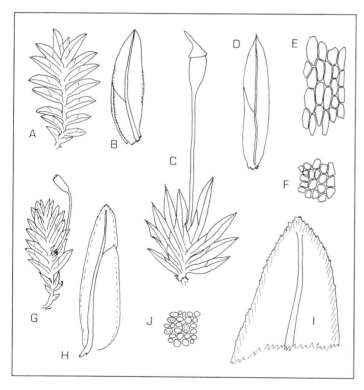

Fig. 15. *A–B:* **Fissidens bryoides,** falcatulus form. *A*, habit, (× 8). *B*, leaf (× 26). *C–F:* **Fissidens kegelianus.** *C*, habit (× 15). *D*, leaf (× 26). *E*, cells of vaginant lamina (× 231). *F*, cells of upper part of leaf (× 231). *G–J:* **Fissidens cristatus.** *G*, habit (× 4). *H*, leaf (× 15). *I*, leaf tip (× 55). *J*, cells of dorsal lamina (× 220).

9. Fissidens kegelianus C.M.
Figure 15, C−F

Plants small, delicate, stems to 3−4 mm tall; leaves to 2 mm long, narrowly lanceolate, widely spreading on the stem so that the fronds are fanlike, bordered all around with a narrow band of elongate cells; cells smooth, those of the vaginant laminae pellucid and conspicuously larger than in rest of leaf. Sporophyte terminal.

> **Habitat:** Soil, rock, rarely wood, in deep shade in low forests.
> **Range:** Southern United States; tropical America.
> **Gulf Coast Distribution:** Florida to south-central Louisiana.

A beautiful plant, *F. kegelianus* is fairly common along the central and eastern portion of the Gulf Coast. It is one of our most attractive mosses as seen under the microscope. The widely spreading leaves, the delicate border all around the leaves, and the enlarged cells of the vaginant laminae make it quite distinctive. Forms of *F. bryoides* also have the leaves bordered all around, but the border is coarser and the leaves lack the enlarged cells in the vaginant laminae; nor do the plants have the distinctive, symmetrical, fanlike appearance of *F. kegelianus*.

10. Fissidens cristatus Wils.
ex Mitt.
Figure 15, G−J

Rather coarse, medium-size plants with stems mostly 1−2 cm tall; leaves about 2 mm long, characteristically curved under at the tips when dry, with a distinct border of pale (but not elongate) cells, usually rather coarsely toothed at the apex; cells thick-walled, bulging, rounded, obscure, here and there bistratose, mostly about 8 μm in diameter. Sporophyte lateral.

> **Habitat:** Rock, soil, tree bases and trunks in forests and ravines.
> **Range:** Over most of the Northern Hemisphere; eastern North America to Minnesota and south to the Gulf.
> **Gulf Coast Distribution:** Florida to Texas.

A very common species in our area, *F. cristatus* seems to prefer somewhat wetter situations than the equally common *F. taxifolius*. Its downwardly curved leaf tips and somewhat glossy appearance allow its recognition in the field. Under the microscope the toothed leaf tip, the border of pale cells, and the scattered areas of bistratose cells are characteristic. Probably the only *Fissidens* with which *F. cristatus* would be likely to be confused is *F. adianthoides*, but in the latter the leaf cells are much larger, the pale border is often lacking or not conspicuous, and the leaves do not have bistratose cells.

11. Fissidens adianthoides Hedw.
Figure 16, A−D

Rather robust plants, mostly about 2 cm tall; leaves tending to curl downward at the tips when dry, about 3−4 mm long, usually coarsely toothed at the tips, sometimes with a border of pale (but not elongated) cells; leaf cells unistratose throughout, angular or rounded, plane to bulging, mostly 10−15 μm diameter. Sporophyte lateral.

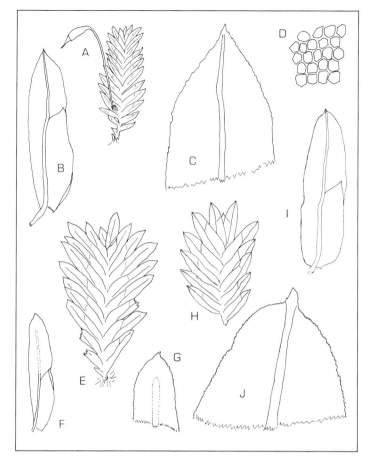

Fig. 16. *A–D:* **Fissidens adianthoides.** *A,* habit (× 3). *B,* leaf (× 15). *C,* leaf tip (× 55). *D,* cells of dorsal lamina (× 220). *E–G:* **Fissidens subbasilaris.** *E,* habit (× 15). *F,* leaf (× 26). *G,* leaf tip (× 55). *H–J:* **Fissidens taxifolius.** *H,* habit (× 5). *I,* leaf (× 26). *J,* leaf tip (× 99).

Habitat: Soil, rock, tree bases, in wet places; frequently on sandstone in streams and on wet cliffs.

Range: Rather widespread in both hemispheres.

Gulf Coast Distribution: Florida to eastern Texas.

The larger leaf cells in one layer only and the common lack of a pale border on the leaves distinguish this species from the much more common *F. cristatus*. As a general rule, plants of *F. adianthoides* are larger and coarser than those of *F. cristatus* and grow in wetter habitats. See comments under the latter name.

Small plants, dull green, stems mostly about 5 mm tall, in loose to dense colonies; leaves about 1 mm long, tending to be downwardly curved at the tips when dry, coarsely toothed above, cells obscure, highly bulging, thick-walled; costa obscure, upper part covered with small cells like those of lamina. Sporophyte lateral.

Habitat: Mostly on tree bark but also on rock, rarely on soil; forests.

12. Fissidens subbasilaris
Hedw.
Figure 16, E–G

Range: North America; Mexico.
Gulf Coast Distribution: Florida to Texas.

A fairly common but usually inconspicuous species, *F. sub-basilaris* resembles a small version of *F. cristatus* in the field. Under the microscope the toothed leaf apex and the costa covered above by small, rounded cells identify it at once. This species is often abundant on the bark of leaning trees, especially white oaks, in mesic woods.

13. **Fissidens taxifolius** Hedw.
Figure 16, H–J

Coarse, dark green plants growing in dense, sordid colonies; stems about 1 cm tall; leaves about 2 mm long, rather abruptly pointed or rounded above and with a strong apiculation; cells rather thick-walled, bulging or faintly unipapillose; margins serrulate, often pale, marginal row of cells slightly differentiated. Sporophytes on very short branches at base of stem.

> **Habitat:** Soil, rocks, tree bases in forests and along ditches, stream banks, and ravines.
> **Range:** Northwestern and eastern North America; Mexico and the Caribbean; Europe; eastern Asia.
> **Gulf Coast Distribution:** Florida to Texas.

Fissidens taxifolius is a weedy moss occurring throughout the Gulf Coast. Its commoness and robustness make it frequently collected. *Fissidens cristatus* is somewhat similar to *F. taxifolius* but has the leaves coarsely toothed above and usually with a conspicuously pale margin. *F. bushii* is also similar but is smaller, its leaves are usually serrate-toothed at the tips, and the cells of its leaves are strongly bulging and irregularly bistratose.

14. **Fissidens bushii** (Card. & Thér.) Card. & Thér.
Figure 17, A–C

Small plants in usually dense sods; stems 3–4 mm tall; leaves about 1 mm long, curved downward at the tips when dry, crenate-serrate on margins, apex with larger, often irregular teeth and apiculate; cells bulging, thick-walled, obscure, bistratose here and there; costa ending in the apiculus or a few cells below. Sporophyte lateral; rarely fruiting in our area.

> **Habitat:** Soil, tree bases, rocks in low forests; often in disturbed sites.
> **Range:** Eastern North America.
> **Gulf Coast Distribution:** Florida to Texas.

These plants often resemble a small *F. cristatus*, and the leaf cells are quite similar to those of *F. cristatus*. Leaves of *F. bushii*, however, lack the pale margins of *F. cristatus* and the plants are much smaller too. *Fissidens bushii* can also be confused with *F. taxifolius*, but plants of that species are generally larger and the leaves have a stout, prominent apiculus which is lacking in *F. bushii*.

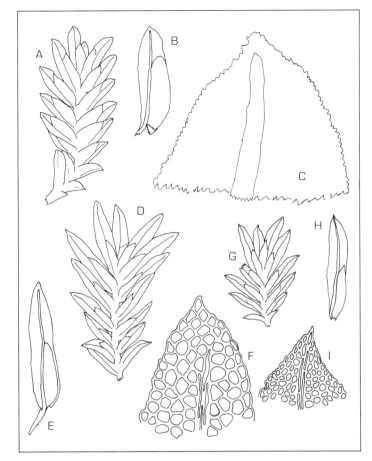

Fig. 17. A–C: **Fissidens bushii.** A, habit (× 15). B, leaf (× 26). C, leaf tip (× 220). D–F: **Fissidens pellucidus.** D, habit (× 15). E, leaf (× 26). F, leaf tip (× 220). G–I: **Fissidens hallii.** G, habit (× 15). H, leaf (× 26). I, leaf tip (× 220).

Small, glossy plants growing in thin colonies; stems 3–4 mm tall; leaves slightly channeled lengthwise when dry, pointed at apex; cells very distinct, thick-walled, 12–14 µm; costa ending a few cells below leaf tip, often appearing to be branched at the tip; leaf margins entire to finely serrate. Sporophyte terminal.

> **Habitat:** Soil and rock in shaded, moist ravines; stream banks; mesic forests.
> **Range:** Southern United States; tropical America.
> **Gulf Coast Distribution:** Florida to Louisiana.

This is a very distinct little moss and rather rare. It has not yet been reported from Texas but should be found in the Big Thicket area eventually. It has been collected in Sabine Parish, Louisiana, just across the Sabine River from Texas. The pointed leaves, large, distinct cells, and costa ending below the leaf tip are characteristic under the microscope. The habitat, glossy appearance, and leaves ± channeled lengthwise identify it in the field. This species is one of the many along the Gulf coastal plain whose main distribution is

15. Fissidens pellucidus
Hornsch.
Figure 17, D–F

to the south, in the tropics, but which have made their way by some means to our shores and have become a regular part of the North American flora. The name *pellucidus* is very appropriate, referring to the large, clear cells of the leaves. These elegant little plants can usually be recognized in the field with a hand lens by the glossy aspect and the slightly longitudinally channeled leaves.

16. **Fissidens hallii** Aust.
Figure 17, G–I

Plants small, in thin colonies, stems about 2 mm tall; leaves 0.5–1 mm long, pointed, usually terminating in a sharp, elongated cell, crenate-serrulate on margins, dentate on margins of vaginant laminae; cells distinct, smooth or with a low, blunt papilla on each, 7–10 μm, on some plants slightly elongated on margins of vaginant laminae; costa ending a few cells below leaf tip or percurrent or slightly excurrent. Sporophyte terminal; apparently rarely fruiting.

> **Habitat:** On tree and stump bases and on clay in low, wet forests.
> **Range:** Southern United States.
> **Gulf Coast Distribution:** Florida, Mississippi, Louisiana, Texas.

This pretty little species is surely present also in Alabama. Its very small size and obscure habitat make it difficult to find. Perhaps the only other *Fissidens* with which *F. hallii* could be confused is *F. bryoides*, forms of which are small and almost lack entirely a border of elongated cells. But even in such extreme forms, the vaginant laminae have the border, and the leaf margins are entire.

17. **Fissidens asplenioides**
Hedw.
Figure 18, A–D

Plants slender, stems to about 2.5 cm tall; leaves 2–3 mm long, curled downward at the tips when dry, broadly acute to rounded at apex, with costa ending well below leaf tip; cells bulging, obscure; leaf margins finely serrate, outer row of cells slightly differentiated. Sporophyte terminal; not fruiting in our area.

> **Habitat:** Moist, shaded stream banks on clay or rock.
> **Range:** Much of the world, mostly in warmer parts.
> **Gulf Coast Distribution:** Known from a few stations in southeastern Louisiana and adjacent Mississippi.

This species was only recently recognized in the United States and is very rare in our area. It could be confused with the much more common *F. polypodioides*, but in the latter species the costa ends right in the leaf tip or 2–3 cells below, the leaf cells are flat, and the leaves are irregularly contorted at the tips when dry rather than being curled. Also, the leaf tips of *F. polypodioides* tend to be more apiculate than in *F. asplenioides*. Careful search in appropriate habitats might well turn up *F. asplenioides* to the east, in other parts of Mississippi, and in Alabama and Florida.

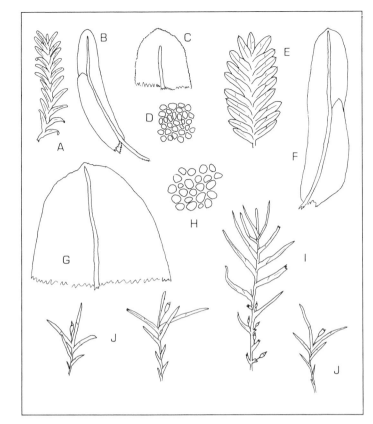

Fig. 18. *A–D*: **Fissidens asplenioides.** *A*, habit (× 3). *B*, leaf (× 15). *C*, leaf tip (× 55). *D*, cells of dorsal lamina (× 220). *E–H*: **Fissidens polypodioides.** *E*, habit (× 3). *F*, leaf (× 15). *G*, leaf tip (× 55). *H*, cells of dorsal lamina (× 220). *I*, **Fissidens fontanus,** habit (× 3). *J*, **Fissidens manateensis,** habit (× 3).

Robust, somewhat glossy plants in dense colonies; stems to 5 cm tall; leaves not much contorted when dry, 3–4 mm long, bluntly rounded at apex or bluntly apiculate; leaf cells plane, distinct; costa ending right in apex or 2–3 cells below; leaf margins essentially entire. Sporophyte lateral on short branches or terminal on major branches; rarely fruiting in our area.

> **Habitat:** Soil on banks of streams and in ravines, occasionally on rock, in forests.
> **Range:** Southern United States, Indiana; tropical America.
> **Gulf Coast Distribution:** Florida to Texas.

Fissidens polypodioides is a handsome moss that often grows in abundance in coastal plain forests. Its large size, glossy appearance, and leaves that are only slightly contorted when dry, make it easy to recognize in the field. The only moss it could be confused with is the very rare *F. asplenioides*, but in that species the leaves are curled at the tips when dry. Under the microscope the flat, distinct cells of the leaves and the ± entire margins of the leaves distinguish *F. polypodioides* from *F. asplenioides*.

18. **Fissidens polypodioides**
Hedw.
Figure 18, E–H

19. Fissidens fontanus
(B.-Pyl.) Steud.
Figure 18, I

Fissidens debilis Schwaegr.

Plants dark green, slender, soft, feathery, aquatic; stems to several cm long; leaves usually remote on stems, 2–4 mm long, linear. Sporophytes borne in axils of lower leaves; peristome teeth truncate, perforate and papillose above.

> **Habitat:** Lakes, bayous, streams, ponds; on rocks, roots, tree bases.
> **Range:** Europe; Africa; North America; Mexico; Cuba; southern South America.
> **Gulf Coast Distribution:** Florida to Texas.

Plants of this species are not often conspicuous in nature because they are dark in color and grow in obscure habitats. Yet they are fairly common and can be found by careful search in appropriate habitats. Most specimens are fertile. The sporophytes are usually clustered in the lower leaf axils. Many collections are taken from exposed tree bases or rocks, indicating that the plants can survive drying for considerable periods when water levels fall.

20. Fissidens manateensis
Grout
Figure 18, J

Similar in many respects to the preceding but generally smaller. Sporophytes borne terminally on major branches; peristome teeth long and slender, obliquely striate above.

> **Habitat:** Lakes, bayous, streams, ponds; on rocks, roots, tree bases.
> **Range:** Southern United States.
> **Gulf Coast Distribution:** Florida, Louisiana.

When fruiting, *F. manateensis* is easily distinguished from *F. fontanus* by the sporophytes terminal on its main branches. Most collections have sporophytes, but they are not conspicuous due to their small size. Sterile specimens are probably not distinguishable from *F. fontanus*. These two species, *F. fontanus* and *F. manateensis*, seem to share the same habitats and probably grow together in many places. The latter species has not been collected in Texas, Mississippi, or Alabama, but almost surely occurs in those states. One Louisiana specimen was found in a water-filled knothole approximately 6 feet up in a Spanish oak (*Quercus falcata* var. *pagodaefolia*).

FISSIDENS DONNELLII Aust. and F. RADICANS Mont. have been reported from our area but probably do not occur here. The former species was reported from Alabama by Wilkes (1965) and from Mississippi by Rogers and Griffin (1974), and the latter was reported from Mississippi by Wilkes (1963) and again by Rogers and Griffin (1974). I have seen most of the specimens upon which these reports were based. Some of them are *F. bushii*, others are *F. ravenelii*, still others are mixtures of both *F. bushii* and *F. ravenelii*. Both *F. donnellii* and *F. radicans* are probably restricted in the United States to southern Florida.

ARCHIDIACEAE Schimp.

Plants very small, yellowish to dark green, in thin to dense, low sods; stems erect, often capitate, simple or branched; leaves increasing in size upwards, erect to erect-spreading, ovate to linear-lanceolate or triangular, margins entire to serrulate or irregularly crenulate, plane or recurved; cells smooth, rectangular to rhombic-hexagonal or rhombic, areolation uniform or irregular; costa single, subpercurrent to strongly excurrent. Cleistocarpous; setae lacking; capsules immersed, globose, up to 750 μm diameter, terminal or lateral; spores large (50–310 μm diameter), mostly few per capsule (4–176).

The Archidiaceae are small, inconspicuous mosses that occur mostly, in our area, on poor, sandy soil. They are not often collected. This treatment is based on that of Snider (1975), who studied the family on a worldwide basis. The family includes only the single genus, *Archidium*. Six species of *Archidium* are known in North America and all of them occur along the Gulf Coast.

ARCHIDIUM Brid.

These obscure little mosses are best recognized by their small size and large spores. The mostly narrow, straight leaves and capitate stems are also helpful in recognizing the genus. Because the capsules are immersed, it is easy to pass over plants of *Archidium* under the assumption that they are sterile plants of other mosses, in the Pottiaceae or Dicranaceae, for example. The large spores are remarkable. A student once said, on seeing the spores of *Archidium* for the first time, that they appeared to be "as big as golf balls." The key to the species of *Archidium* was prepared by Dr. Jerry A. Snider.

Special Study Methods for *Archidium*

The detailed examination of *Archidium* plants required to determine the sexual condition is difficult in ordinary water mounts. If the plants are first soaked in water and then placed in Hoyer's solution (see p. 19) it will be much easier to make the dissection and find the disposition of the sex organs.

1. Plants synoicous; costa of perichaetial leaves weak, ending below apex to subpercurrent; median cells of perichaetial leaves loosely long-rectangular   1. A. MINUS
1. Plants autoicous or paroicous; costa of perichaetial leaves distinct, obviously percurrent to strongly excurrent; median cells of perichaetial leaves uniformly rhombic-hexagonal, short-rectangular, or prosenchymatous, or median cells irregular, containing a mixture of quadrate, trapezoidal, rhombic-hexagonal, rhomboidal and rectangular cells   2

2. Median cells of perichaetial and upper stem leaves forming an obvious irregular areolation, cells varied from quadrate to short-rectangular to trapezoidal to rhomboidal within a single leaf; costa short-excurrent to strongly excurrent into a terete, pellucid, minutely-spined hair-point; autoicous, antheridial buds terminal on short lateral branches .... 6. A. DONNELLII

2. Median cells of perichaetial and upper stem leaves forming a uniform areolation, cells rhombic-hexagonal, or elongate- or linear-rhomboidal, or prosenchymatous; costa percurrent to strongly excurrent; plants autoicous or paroicous 3

3. Median cells of perichaetial leaves mostly less than 15 μm (9–14 μm) wide 4

3. Median cells of perichaetial leaves mostly more than 15 μm (16–34 μm) wide 5

4. Plants paroicous, antheridia naked or enclosed by 1–2 bracts in axils of perichaetial leaves .... 2. A. ALTERNIFOLIUM

4. Plants autoicous, antheridial and archegonial buds lateral, subsessile to sessile in axils of stem leaves, or archegonia terminal and antheridia in axillary buds, or antheridia and archegonia terminal on separate branches arising from older, decumbent and stoloniferous stems (the plants thus appearing to be dioicous) .... 4. A. OHIOENSE

5. Plants paroicous, antheridia naked in axils of perichaetial leaves, or occasionally in small bracts or naked in axils of uppermost stem leaves .... 3. A. TENERRIMUM

5. Plants autoicous, antheridia terminal on short to sessile lateral branches, or antheridial and archegonial buds terminal on separate branches arising from older, decumbent and stoloniferous stems (the plants thus appearing to be dioicous) .... 5. A. HALLII

1. **Archidium minus** (Ren. & Card.) Snider
Figure 19, A–B

Plants mostly 1–3 mm tall; stems mostly simple; median cells of upper stem leaves rectangular, 35–46 × 15–20 μm; costa weak, ending below apex to subpercurrent; antheridia and archegonia intermixed, terminal; perichaetial leaves larger than stem leaves. Capsules terminal, 1–3 per stem, to 650 μm diameter; spores 16–48 per capsule, 119–173 μm diameter, smooth to densely papillose.

Habitat: Sandy soil on banks and roadsides.
Range: Southern United States.
Gulf Coast Distribution: Louisiana.

This little moss is apparently known in our area only from a single collection from St. Tammany Parish, Louisiana. The specimen was collected by A. B. Langlois in 1892 and was used as the basis for the original description of this species. *Archidium minus* also is known from peninsular Florida and southeastern Georgia. The delicate little plants of *A. minus* are strongly reminiscent of the Ephemeraceae.

Plants to 20 mm tall, in dense sods; stems simple or branched; median cells of upper stem leaves 45–160 × 9–12 μm; alar cells quadrate to short-rectangular, in at least 2–3 rows on basal margins; costa percurrent to short-excurrent; paroicous, antheridia naked in axils of perichaetial leaves (rarely enclosed by 1–2 bracts); perichaetia terminal but often appearing lateral due to innovations; perichaetial leaves larger than stem leaves. Capsules terminal or (apparently) lateral, to 750 μm diameter; spores 8–36 per capsule, 127–262 μm diameter, smooth to faintly papillose.

Habitat: Sandy soil in overflow areas, roadside and river banks, around temporary ponds.

Range: Eastern United States; Europe; northern Africa.

Gulf Coast Distribution: Florida to eastern Texas.

The paroicous sexual condition and relatively narrow leaf cells are distinctive; the more common *A. tenerrimum* is also paroicous but mostly has much wider leaf cells.

2. **Archidium alternifolium**
(Hedw.) Schimp.
Figure 19, C–D

Archidium longifolium Lesq. & James

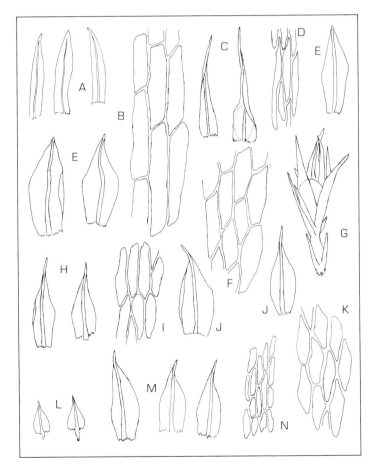

Fig. 19. *A–B*: **Archidium minus.** *A*, leaves (× 15). *B*, median cells (× 231). *C–D*: **Archidium alternifolium.** *C*, leaves (× 15). *D*, median cells (× 231). *E–F*: **Archidium tenerrimum.** *E*, leaves (× 15). *F*, median cells (× 231). *G–I*: **Archidium ohioense.** *G*, habit (× 15). *H*, leaves (× 15). *I*, median cells (× 231). *J–K*: **Archidium hallii.** *J*, leaves (× 15). *K*, median cells (× 231). *L–N*: **Archidium donnellii.** *L*, leaves of sterile stem (× 15). *M*, perichaetial leaves (× 15). *N*, median cells from perichaetial leaf (× 231).

3. **Archidium tenerrimum**
Mitt.
Figure 19, E–F

Plants to 10 mm tall, in thin to dense mats or sods; stems mostly simple; median cells of stem leaves variable in size, 16–34 μm wide; alar cells quadrate to short-rectangular; costa percurrent to short-excurrent; paroicous, antheridia naked in axils of outer perichaetial leaves; perichaetia terminal; perichaetial leaves larger than stem leaves. Capsules terminal, 470–790 μm diameter; spores 4–48 per capsule, 128–218 μm diameter, smooth.

> **Habitat:** Sandy banks; roadsides; forest edges.
> **Range:** Southeastern United States.
> **Gulf Coast Distribution:** Florida to southern Texas.

Apparently fairly common along the Gulf Coast, especially on roadsides. See comments under *A. alternifolium.*

4. **Archidium ohioense**
Schimp. ex C.M.
Figure 19, G–I

Archidium floridanum Aust.

Plants to 20 mm tall; stems mostly simple; median cells rhomboidal to ± prosenchymatous; alar cells short-rectangular to quadrate, in 2–6 rows along basal margins; costa percurrent to excurrent into a hair-point. Autoicous; antheridial buds lateral and sessile, and terminal, or terminal on short branches; perichaetial leaves variable; costa percurrent to strongly excurrent, often ending in a hair-point. Capsules 270–600 μm diameter; spores 4–60 per capsule, 110–310 μm diameter, smooth to densely papillose.

> **Habitat:** Soil, usually in moist situations.
> **Range:** Eastern North America; Africa; Asia.
> **Gulf Coast Distribution:** Louisiana; eastern Texas.

See comments under *A. hallii.*

5. **Archidium hallii** Aust.
Figure 19, J–K

Plants to 8 mm tall; stems mostly branched; median cells rhombic-hexagonal to rhomboidal; alar cells short-rectangular to quadrate, in 2–6 rows along basal margins; costa percurrent to slightly excurrent; autoicous; antheridial buds terminal on short lateral branches or on longer branches arising from old stems; costae of perichaetial leaves percurrent to short-excurrent. Capsules 500–650 μm diameter; spores 8–36 per capsule, 143–253 μm diameter, smooth to granular.

> **Habitat:** Sandy soil on banks and along roads.
> **Range:** Southeastern United States; Mexico; southern South America.
> **Gulf Coast Distribution:** Eastern Texas.

This species is only rarely collected. Its larger perichaetial leaves, with wider cells, should separate it from *A. ohioense.*

6. **Archidium donnellii** Aust.
Figure 19, L–N

Plants to 8 mm tall; stems branching; upper stem leaves with margins recurved; median cells irregular in shape, basal cells not much different in size from median cells; costa percurrent to short-excurrent; autoicous, antheridial buds terminating short branches;

perichaetial leaves larger than stem leaves, costa short-excurrent to strongly excurrent into a hair-point. Capsules 425–700 μm diameter; spores 4–60 per capsule, 125–230 μm diameter, smooth to granulose.

Habitat: Sandy soil on road banks, ditches; around fields.
Range: Eastern United States.
Gulf Coast Distribution: Eastern Texas.

Apparently known in our area only from Bastrop County, Texas. The irregular leaf cells and ± undifferentiated basal cells should distinguish *A. donnellii* from *A. ohioense* and *A. hallii*.

DITRICHACEAE Limpr.

Plants tiny to medium-size, in tufts or sods, green to yellowish green, sometimes tinged with red; stems ± erect, mostly simple; leaves mostly lanceolate, acuminate to subulate; margins erect to recurved; median cells quadrate to linear, smooth (rarely papillose-roughened), lower cells longer, sometimes lax, not differentiated at basal angles; costa single, subpercurrent to long-excurrent. Setae very short to elongate; capsules immersed and lacking differentiated operculum and peristome or exserted and with operculum and peristome, globose to cylindrical, erect to inclined, sometimes asymmetric; peristome single, of 16 teeth usually split to base; calyptra mostly cucullate, rarely mitrate.

The Ditrichaceae are mostly soil-dwelling mosses of a weedy nature, occupying ± disturbed sites.

Plants tiny, cleistocarpous, capsules immersed	1. Pleuridium
Plants larger, capsules operculate and exserted	2. Ditrichum

Plants very small, yellowish green to green or brownish, loosely clustered to scattered in thin or dense, sometimes grasslike, mats; stems short, erect, mostly simple; leaves mostly narrow and subulate from a broader base, straight and erect to erect-spreading or appressed when dry; cells smooth, short to elongate in upper part of leaf, more lax below, not differentiated in alar regions; costa mostly filling the acumination and excurrent, at least in the upper and perichaetial leaves. Cleistocarpous; setae very short, straight or curved; capsules immersed, apiculate to ± short-rostrate; calyptra small, mostly cucullate, sometimes mitrate.

Pleuridium is a genus of very small, annual mosses of disturbed habitats such as old fields, floodplains, roadsides, and pond and stream edges. Along the Gulf Coast these mosses are most com-

1. PLEURIDIUM Rabenh.

monly found with sporophytes from March to June. Four species have been reported from our area.

1. Calyptra mitrate; stomata near midcapsule	4. P. PALUSTRE
1. Calyptra cucullate; stomata at base of capsule	2
2. Slender, julaceus, sterile stems usually present and abundant; awn of perichaetial leaves much less than ½ leaf length; leaves appressed	3. P. SULLIVANTII
2. Sterile, julaceus stems absent; awn of perichaetial leaves ± equal to or much longer than leaf length; leaves erect-spreading	3
3. Leaves with long, slender acumination, the acumination much longer than the broad base; upper and perichaetial leaves 3–4 mm long; plants grasslike in appearance	1. P. SUBULATUM
3. Leaves mostly with shorter acumination, the acumination mostly ± equal to or shorter than the broad base; upper and perichaetial leaves 2–3 mm long; plants not grasslike in appearance	2. P. RAVENELII

1. **Pleuridium subulatum**
(Hedw.) Rabenh.
Figure 20, A

Plants often grasslike in appearance, in tufts or mats; leaves long and slenderly subulate-acuminate from an oblong to ovate base, the acumination usually much longer than the broad basal portion of the leaf; upper and perichaetial leaves 3–4 mm long. Setae very short; capsules bluntly apiculate, about 1 mm long.

Habitat: Disturbed sites in old fields; roadside banks.
Range: Europe; northern Africa; Asia; eastern North America.
Gulf Coast Distribution: Florida to Texas.

The grasslike appearance of the colonies and the long and slenderly subulate leaves make this species easy to recognize.

2. **Pleuridium ravenelii** Aust.
Figure 20, B

Plants in low, thin or dense mats; leaves subulate-acuminate from an oblong to ovate base, the acumination mostly equal to or shorter than the basal portion (sometimes longer); upper and perichaetial leaves 2–3 mm long. Setae very short; capsules bluntly apiculate, mostly about 0.75 mm long.

Habitat: Old fields, pastures, open forests, pond margins, along streams.
Range: Eastern United States.
Gulf Coast Distribution: Louisiana and eastern Texas.

This species has not yet been reported from Florida, Alabama, or Mississippi, but it probably occurs throughout the Gulf Coast. It is similar in some respects to *P. subulatum* but differs in the shorter leaves which usually have the subulation shorter than or equal to the broad base. Crum and Anderson (1981) note that in this species the leaf shoulders, at the base of the acumination, often have jagged interruptions in the margins, a feature lacking in *P. subulatum*.

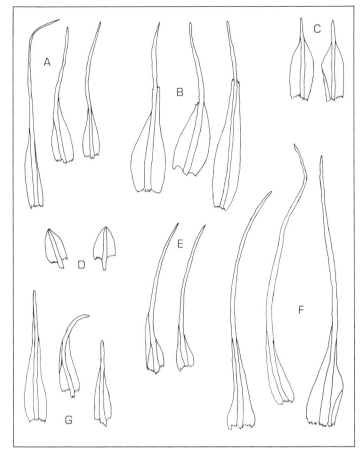

Fig. 20. *A,* **Pleuridium subulatum,** leaves (× 15). *B,* **Pleuridium rav-enelii,** perichaetial leaves (× 15). *C–D:* **Pleuridium sullivantii.** *C,* perichaetial leaves (× 15). *D,* leaves from sterile stem (× 15). *E,* **Pleuridium palustre,** perichaetial leaves (× 15). *F,* **Ditrichum pal-lidum,** leaves (× 15). *G,* **Ditrichum pusillum,** leaves (× 15).

Plants in thin mats, usually with abundant, slender, erect, sterile, julaceus stems intermixed with the fertile stems; leaves appressed, subulate from an oblong base, acumination shorter than the basal part; upper and perichaetial leaves 1.5–2 mm long. Setae very short; capsules bluntly apiculate, about 0.75 mm long.

3. Pleuridium sullivantii Aust.
Figure 20, C–D

    Habitat: Disturbed sites in open forests; along streams.
    Range: Eastern United States.
    Gulf Coast Distribution: Louisiana; eastern Texas.

This rare little moss is distinctive because of its appressed leaves and sterile, julaceus stems intermixed with the fertile ones. It is known in our area only from two collections from near St. Fran-cisville, in Louisiana, and from a few counties in eastern Texas.

Plants in rather dense, low mats; leaves long and slenderly subulate-acuminate from an ovate to oblong base, acumination several times longer than the leaf base; upper and perichaetial leaves 2–4 mm long. Setae short and rather fleshy; capsules (including apiculation) about 1 mm long, bluntly to stoutly apiculate to almost rostrate; calyptra mitrate.

4. Pleuridium palustre (Bruch & Schimp.) B.S.G.
Figure 20, E

Habitat: Floodplain forests; pond margins; on soil.
Range: Europe; eastern United States.
Gulf Coast Distribution: Louisiana.

This moss is apparently very rare in our area; it is known along the Gulf Coast only from a single collection from Tangipahoa Parish, Louisiana. It is similar to *P. subulatum* in having long-subulate upper and perichaetial leaves, but differs in the mitrate calyptra, stomata near the middle of the capsule, and the stoutly apiculate capsule.

2. DITRICHUM Hampe

Plants small to medium-size, in loose tufts or dense sods; stems mostly simple, ± erect; leaves mostly acuminate or subulate from a broader base, often curved, sometimes secund, lamina sometimes absent above shoulders with upper part of leaf then consisting of the excurrent costa alone; margins erect to recurved, sometimes thickened; upper cells subquadrate to linear, smooth (rarely ± papillose), lower cells oblong to rectangular or linear, sometimes ± lax; costa subpercurrent to long excurrent. Setae elongate; capsules cylindric, erect to inclined, sometimes curved, often sulcate when dry and empty; peristome single, the 16 teeth mostly split to the base into two equal, slender, papillose divisions, ± straight to slightly twisted; operculum bluntly conic to conic-rostrate; calyptra cucullate.

The genus *Ditrichum* is best recognized by the habit of the plants and by the characteristic peristome. *Dicranella* is similar but has a different type of peristome. Three species of *Ditrichum* have been reported for the Gulf Coast but probably only two actually occur here. One, *D. pallidum*, is common and abundant.

Seta bright yellow, to 4 cm long; upper leaves 3–5 mm long, margins erect	1. D. PALLIDUM
Seta red-brown, to 1.5 cm long; upper leaves 2–3 mm long, margins recurved	2. D. PUSILLUM

1. Ditrichum pallidum (Hedw.) Hampe
Figure 20, F

Ditrichum currituckI Grout

Plants ± glossy, green to yellowish green, in dense, silky tufts or sods; stems ± erect, mostly simple; leaves erect-spreading, often secund, slenderly long-acuminate-subulate from an oblong base, channeled above, acumination consisting of the excurrent costa; margins mostly entire below, serrulate toward apex; basal cells subquadrate to oblong or rectangular, often ± lax, upper cells linear; costa long-excurrent, composing most of leaf above base. Setae bright yellow; capsules brown, erect to inclined, cylindric, sulcate when dry and empty; peristome less than 1 mm long, not or only slightly twisted; operculum bluntly conic.

Habitat: Open forests, fields, roadsides, lawns; on soil, mostly in somewhat disturbed sites.

Range: Europe; Africa; Japan; eastern North America.
Gulf Coast Distribution: Florida to Texas.

Ditrichum pallidum is one of our commonest mosses; its colonies are very conspicuous in the spring and early summer as soft, shiny, green cushions which become bright yellow as the setae mature. Sterile plants of this moss could be taken for *Campylopus*; see comments under that genus for the differentiation.

Plants small, yellowish to dark green, in thin to dense, low sods; stems erect, mostly simple; leaves mostly erect, rather strict, not or little contorted, lanceolate-subulate, channeled above, margins ± recurved, at least at midleaf, serrulate, irregularly thickened; basal cells rectangular, upper cells subquadrate to rectangular or short-linear; costa distinct to leaf tip, subpercurrent. Setae red to red-brown, at least in lower part; capsules brown, short-cylindric, erect, not sulcate; peristome shorter than in *D. pallidum*; operculum bluntly conic-rostrate.

2. **Ditrichum pusillum** (Hedw.) Hampe
Figure 20, G

Habitat: Bare soil along stream banks and ditches.
Range: Europe; India; northern Africa; North America.
Gulf Coast Distribution: Florida to eastern Texas.

Ditrichum pusillum is a rather nondescript little moss that is apparently quite rare along the Gulf Coast. It is easily overlooked.

DITRICHUM RHYNCHOSTEGIUM Kindb. (*D. henryi* Crum & Anderson) has been reported from Louisiana but probably does not occur on the Gulf Coast (Anderson and Bryan 1958). It is similar to *D. pallidum* but differs, among other ways, in its orange-yellow seta and longer peristome.

CERATODON PURPUREUS (Hedw.) Brid. has been reported from our area (Perry County, Mississippi; Rogers and Griffin 1974), but I have not been able to see the specimen and doubt that this species really occurs along the Gulf Coast. Although this species has a very wide distribution around the world, it is apparently absent in the Gulf South.

DICRANACEAE Schimp.

Plants very small to robust, green to yellowish green or white, mostly in dense tufts, often densely radiculose below; stems erect, mostly simple; leaves often falcate-secund, mostly narrow and lanceolate, often subulate, sometimes several cell layers thick; upper cells short to elongate, mostly smooth, lower cells mostly larger and elongate, often conspicuously differentiated in alar regions; costa single, usually strong, sometimes very broad and comprising most of the leaf, ending near leaf apex to strongly excurrent. Sporophytes mostly with operculum and peristome, sometimes cleistocarpous; setae mostly elongate, sometimes very short; capsules erect to inclined, often curved and asymmetric, frequently ribbed when dry, neck sometimes differentiated; peristome single, teeth usually split ± to the middle; operculum usually obliquely rostrate; calyptra mostly cucullate, sometimes mitrate.

We have six genera of this large family in our area.

1. Plants pale, often whitish or whitish green; leaves fleshy, several cell layers thick	6. Leucobryum
1. Plants green to yellowish green; leaves (except the costae) one cell layer thick	2
2. Plants very small, cleistocarpous; capsules mostly immersed	1. Bruchia
2. Plants small to large, stegocarpous; capsules exserted	3
3. Capsules with prominent, long, slender neck	2. Trematodon
3. Capsules lacking long, slender neck	4
4. Alar cells conspicuously differentiated, at least in some leaves; plants rarely or not producing sporophytes in our area	5
4. Alar cells not conspicuously differentiated; plants commonly producing sporophytes	3. Dicranella
5. Costa very broad, ⅓ or more width of leaf base	4. Campylopus
5. Costa narrower, less than ⅓ width of leaf base	5. Dicranum

1. Bruchia Schwaegr.

Plants mostly very small, in thin, yellowish green colonies; stems erect, mostly short and simple; leaves crowded at stem tips, mostly linear-subulate from a broader base, the upper portion often composed mostly of the costa, sometimes broader, spreading to ascending, straight or flexuous when dry, rarely imbricate, margins entire to serrulate, sometimes recurved; cells smooth or papillose, quadrate to short-rectangular or linear above, broader in basal portion; costa percurrent to long-excurrent. Cleistocarpous; usually producing abundant sporophytes; setae mostly short but sometimes elongate; capsules apiculate, immersed to exserted, with prominent, thick neck, upper portion of capsule often bright yellow-orange or reddish; spores pitted, papillose, spinose, or reticulate; calyptra mitrate, smooth or papillose.

Although plants of *Bruchia* are mostly very small, the colonies are usually conspicuous due to the abundance of the often brightly colored sporophytes, which mature from late winter to spring. It is essential to have mature spores for identification. Six species of *Bruchia* occur along the Gulf Coast; only rarely are any of them very abundant.

1. Calyptra rough-papillose to spinose	2
1. Calyptra essentially smooth	3
2. Spores reticulate; calyptra prominently papillose to spinose	3. B. RAVENELII
2. Spores pitted; calyptra with low papillae	4. B. CAROLINAE
3. Spores pitted	5. B. BREVIFOLIA
3. Spores papillose, spinose, or reticulate	4
4. Leaves broad, apex abruptly short-acuminate; stems elongate	6. B. HALLII
4. Leaves, at least upper and perichaetial, subulate	5
5. Perichaetial leaves mostly exceeding capsules; capsules truncate at base, brown-orange to red above	2. B. DRUMMONDII
5. Perichaetial leaves mostly not exceeding capsules; capsules gradually tapering to seta at base, yellow to brown above	1. B. FLEXUOSA

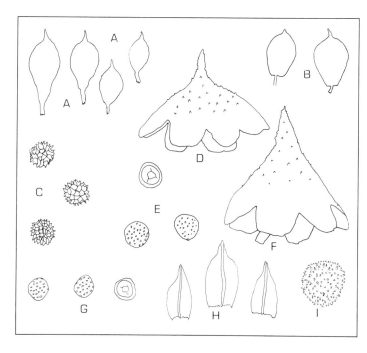

Fig. 21. *A*, **Bruchia flexuosa,** capsules (× 15). *B*, **Bruchia drummondii,** capsules (× 15). *C–D*: **Bruchia ravenelii.** *C*, spores (× 231). *D*, calyptra (× 55). *E–F*: **Bruchia carolinae.** *E*, spores (× 231). *F*, calyptra (× 55). *G*, **Bruchia brevifolia,** spores (× 231). *H–I*: **Bruchia hallii.** *H*, leaves (× 23). *I*, spore (× 231).

Plants mostly very small and short-stemmed; leaves straight or ± flexuous, the subula essentially smooth to distinctly serrulate and papillose-roughened. Setae very short to slightly elongate, sometimes curved; capsules mostly ± exserted, sometimes immersed, brown to yellow above; spores papillose to spinose or reticulate; calyptra smooth.

Habitat: Old fields; roadsides, ditch banks.
Range: Eastern North America.
Gulf Coast Distribution: Florida to Texas.

Bruchia flexuosa is quite variable in most respects. The species with which it is most likely to be confused is *B. drummondii.* In

1. **Bruchia flexuosa** (Sw. ex Schwaegr.) C.M.
Figure 21, A

Bruchia texana Aust.
B. sullivantii Aust.
B. donnellii Aust.

that species the spores are clearly reticulate with large areolae, and the capsules are usually bright orange to red above, rather than brown to yellow. The status of *B. texana*, *B. sullivantii*, and *B. donnellii* is uncertain, and in this work they are all treated under the concept of *B. flexuosa*. See comments under *B. drummondii*.

2. Bruchia drummondii
Hampe ex E.G. Britt.
Figure 21, B

Plants very small, almost stemless; leaves subulate, the subula channeled and smooth or slightly roughened. Setae very short, shorter than the capsules; capsules usually immersed, bright orange to red above when mature; spores with large clear reticulations about 8 μm diameter; calyptra smooth.

> **Habitat:** Sandy soil; banks, fields, roadsides, mostly in pine areas.
> **Range:** Southeastern United States.
> **Gulf Coast Distribution:** Florida to eastern Texas.

This moss, with its orange to red capsules, leaves usually exceeding the capsules, and its spores with large clear reticulations, is quite distinctive. Its capsules are noticeably truncate at the base, a feature that aids in distinguishing it from *B. flexuosa*.

3. Bruchia ravenelii Wils.
ex Sull.
Figure 21, C–D

Plants virtually stemless; leaves subulate, the subula ± smooth to somewhat roughened. Setae usually shorter than the capsules; capsules mostly immersed, yellow to bright orange or reddish above; spores reticulate; calyptra with low to spinose papillae on the sides and beak.

> **Habitat:** Sandy soil; banks and clearings in pine areas.
> **Range:** Southeastern United States.
> **Gulf Coast Distribution:** Florida; Alabama; Louisiana; eastern Texas.

The papillose calyptra and reticulate spores are diagnostic. *Bruchia carolinae* also has the calyptra roughened, but to a lesser degree, and its spores are pitted rather than reticulate.

4. Bruchia carolinae Aust.
Figure 21, E–F

Similar in most respects to *B. ravenelii*; differing primarily in the pitted, rather than reticulate, spores, and in the smaller papillae on the calyptra. In *B. ravenelii* the calyptra has rather prominent, sometimes spinose papillae. The capsule of *B. carolinae* is yellow to brownish above, rather than orange-red as it often is in *B. ravenelii*.

> **Habitat:** Sandy soil; roadsides and old fields.
> **Range:** Southeastern United States.
> **Gulf Coast Distribution:** Florida and Louisiana.

Probably present all along the Gulf Coast; see comments under *B. brevifolia*.

Plants very small, essentially stemless; leaves short, with short acumination, reaching the lower part of the capsule or shorter. Setae very short, shorter than the capsules; capsules exserted, bright orange above when mature; spores pitted; calyptra smooth.

Habitat: Sandy soil; banks and fields.
Range: Southeastern United States.
Gulf Coast Distribution: Eastern Texas.

This species is apparently rare. The short leaves and pitted spores are distinctive. Although it has been reported from Louisiana, I have not seen material from there; *B. carolinae*, our other species with pitted spores, has a papillose calyptra.

Plants with ± elongate stems; leaves imbricate-appressed when dry, ovate-lanceolate, abruptly short-acuminate, not subulate, entire. Capsules exserted, brown; spores spinose; calyptra smooth.

Habitat: Sandy soil in old fields.
Range: Southeastern United States.
Gulf Coast Distribution: Eastern Texas.

Bruchia hallii is a very rare species; it is known only from a few collections from Texas, Arkansas, and North Carolina. The imbricate, rather broad, nonsubulate leaves and spinose spores are distinctive.

BRUCHIA FUSCA E.G. Britt. was reported from Smith County, Texas, by Whitehouse and McAllister (1954). However, I have examined the specimen upon which the report was based (University of Iowa Herbarium), and it is actually *B. hallii* Aust.

Plants in thin, low, yellowish green sods; stems very short, erect; leaves flexuous when dry, subulate-acuminate from a broader, clasping base, upper part concave, median and upper cells smooth, quadrate to short-rectangular, cells of basal portion much larger, rectangular to rhombic, often yellow; margins erect or ± recurved, entire; costa slender, ending in apex to slightly excurrent. Setae very long and slender, yellow; capsules cylindric, curved, each with a conspicuous neck much longer than the capsule itself; peristome single, of 16 teeth, the teeth vertically striate and perforated to ± divided; operculum slenderly and slightly obliquely long-rostrate; calyptra cucullate, naked.

Habitat: Bare soil on banks, along paths; often in fields and other disturbed sites.
Range: Europe; Asia; Australia; eastern United States; Central America; Caribbean; South America.
Gulf Coast Distribution: Florida to eastern Texas.

This species is easy to recognize when sporophytes are present, due to the strikingly long necks on the capsules. It is fairly common

5. **Bruchia brevifolia** Sull.
Figure 21, G

6. **Bruchia hallii** Aust.
Figure 21, H–I

2. TREMATODON Michx.

Trematodon longicollis Michx.
Figure 22, A–B

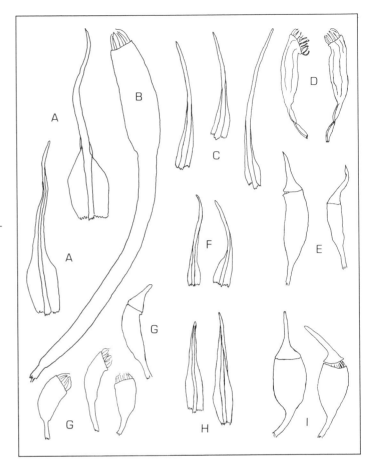

Fig. 22. *A–B:* **Trematodon longicollis.** *A,* leaves (× 15). *B,* capsule (× 15). *C–E:* **Dicranella heteromalla.** *C,* leaves (× 15). *D,* dry capsules (× 15). *E,* soaked capsules (× 15). *F–G:* **Dicranella varia.** *F,* leaves (× 15). *G,* capsules (× 15). *H–I:* **Dicranella hilariana.** *H,* leaves (× 15). *I,* capsules (× 15).

in our area, with sporophytes produced from late winter to June or July. The usually massed, yellow setae make colonies of this moss very conspicuous.

3. DICRANELLA (C.M.) Schimp.

Plants small to medium-size, dull green to yellowish green; scattered to gregarious or tufted; stems erect, simple or forked; leaves erect to flexuous-spreading, often secund, lanceolate to subulate from a broader base; cells quadrate to rectangular or short-linear, smooth; margins plane to recurved, entire below, entire to serrulate above; costa strong, percurrent to excurrent. Setae elongate; capsules ovoid to short-cylindric, erect and symmetric to inclined and asymmetric; peristome teeth divided to about the middle into slender papillose segments; annulus present or lacking; operculum obliquely rostrate.

Plants of this genus sometimes resemble small plants of *Dicranum* sp., hence the name *Dicranella*, meaning "little *Dicranum*." The gametophytes are quite variable, in general, and it is probably futile to attempt to identify sterile specimens to the species level. However, *Dicranella* species produce sporophytes freely, although sometimes sparingly, over much of the year along the Gulf Coast. Three species occur with some abundance in our area.

1. Capsules erect and symmetric, smooth when dry and empty; annulus present and dehiscent; seta yellow — 3. D. HILARIANA
1. Capsules ± inclined and asymmetric, smooth or furrowed when dry and empty; annulus lacking or poorly developed and persistent on the urn; seta yellow or red — 2
  2. Seta yellow; capsule furrowed when dry and empty; annulus present and persistent but poorly developed; operculum slenderly rostrate — 1. D. HETEROMALLA
  2. Seta red; capsule smooth when dry and empty; annulus lacking; operculum bluntly rostrate — 2. D. VARIA

Plants small to medium-size; leaves mostly conspicuously secund but sometimes ± straight and erect, subulate-setaceous from a slightly broadened base, the subula roughened above and consisting mostly of the costa; lamina disappearing upwards along the costa, its cells rectangular, basal cells often yellow at the insertion; margins plane, serrulate above; costa very strong, broad below, long-excurrent. Setae yellow; capsules short-cylindric, curved, asymmetric and furrowed when dry, usually with a distinct depression just below the strongly oblique mouth; annulus ± differentiated but not dehiscent, remaining attached to the urn; spores finely granular; operculum slenderly rostrate.

1. Dicranella heteromalla (Hedw.) Schimp.
Figure 22, C–E

  Habitat: Banks of ravines and streams; in forests.
  Range: Circumpolar; widespread in North America; northern South America.
  Gulf Coast Distribution: Florida to eastern Texas.

Plants of this species are generally more robust than those of the other two Gulf Coast species of *Dicranella*, and the leaves are more consistently secund. The elongate, furrowed capsules with oblique mouths make this species distinctive. It is widespread and rather common in our area.

Plants small, yellowish green; leaves straight, erect, sometimes a little secund, acuminate to subulate from broader base; median cells rectangular, becoming larger and lax below; margins mostly entire, sometimes denticulate above and at apex, usually recurved at least in part; costa percurrent to excurrent, usually comprising most of the upper part of the leaf. Setae red; capsules short, mostly oblique and slightly to conspicuously asymmetric; annulus lacking; spores nearly smooth to granular; operculum bluntly rostrate.

2. Dicranella varia (Hedw.) Schimp.
Figure 22, F–G

  Habitat: Soil and rock, including limestone; roadsides, banks, ravines.

Range: Circumpolar; widespread in North America, south to the Caribbean and Central America.
Gulf Coast Distribution: Florida to eastern Texas.

This species is widespread but not very common along the Gulf Coast. The red seta, lack of an annulus, and short, oblique, smooth capsule are distinctive.

3. **Dicranella hilariana** (Mont.) Mitt.
Figure 22, H–I
Dicranella herminieri Besch.

Plants tiny, yellowish green; leaves erect or somewhat spreading, straight or flexuous, sometimes slightly secund, lanceolate-acuminate from a broader base, apex acute and entire to bluntly rounded and toothed by projecting cells; median and upper cells quadrate to rectangular or linear, larger and lax below; margins entire below, entire above or dentate towards apex, mostly recurved at least in part; costa ending in or just below leaf tip. Setae yellow; capsules short-ovoid, erect and symmetric; annulus present and dehiscent; spores granular; operculum slenderly rostrate.

Habitat: Sandy or clay banks along streams, paths, ditches, roadsides; in forests.
Range: Southeastern United States; tropical America.
Gulf Coast Distribution: Florida to Texas.

This little moss is fairly common essentially throughout our area. The erect, symmetric capsules with well-developed, dehiscent annulus make it distinct among our other species of *Dicranella*.

Dicranella rufescens (With.) Schimp., similar to *D. hilariana* but lacking an annulus, has been reported from Louisiana (Taylor 1927). However, it has a more northern range and probably does not occur in our area. I was unable to locate the specimen upon which Taylor's report was based.

4. **CAMPYLOPUS Brid.**

Plants small and slender to rather robust, yellowish green to green or darker, in thin to dense tufts or sods, often with the leaves clustered at the stem tips and forming heads, usually matted with tomentum below; stems erect, simple or forked; leaves mostly narrow and acuminate or subulate, erect-appressed to spreading, straight or flexuous; median cells quadrate to rhombic or elongate, lower cells elongate and larger, cells in alar regions often inflated and colored in conspicuous groups, sometimes forming auricles; costa very strong and wide, filling most of the upper portion of the leaf, percurrent to long excurrent, sometimes forming a hyaline hair-point. Not producing sporophytes in our area.

Although *Campylopus* plants are always sterile in our area they are easy to recognize, at least under the microscope, by the very broad costa and the often differentiated alar cells. Sterile *Ditrichum pallidum* could be taken for a *Campylopus*; however, the greater portion of the *Ditrichum* leaf is composed of the multilayered costa, with very long and narrow cells, while in *Campylopus* there are always unistratose wings of lamina along the costa, composed of quadrate to shortly linear cells.

Four species of *Campylopus* occur in our area, mainly in the eastern part.

Although *Campylopus* does not produce sporophytes in our area, its method of exposing its spores for dispersal is remarkable in some species and deserves mention. In certain species of the genus the seta curves during development and drives the capsule down into the upper leaf axils, where the calyptra with the contained operculum is caught fast. Later, the seta straightens somewhat, pulling the capsule away from the trapped calyptra and operculum, and the spores are thus exposed for dissemination.

1. Leaves mostly tightly appressed to stems when dry, with conspicuous hyaline tips; costa with dorsal lamellae — 1. C. PILIFER
1. Leaves mostly variously flexuous-spreading, hyaline tips lacking or inconspicuous; costa smooth dorsally or with low ridges — 2
 2. Median and upper leaf cells (exclusive of the costa) linear-flexuous — 4. C. ANGUSTIRETIS
 2. Median and upper leaf cells quadrate to short-rectangular or rhombic — 3
3. Median and upper leaf cells rhombic, in rows diverging from the costa; plants usually capitate — 3. C. SURINAMENSIS
3. Median and upper cells quadrate to short-rectangular, in rows parallel to the costa; plants not particularly capitate — 2. C. TALLULENSIS

Plants dark to yellowish green, sometimes blackish, in stiff tufts or sods, tomentum red; leaves appressed or ascending, 3–5 mm long, usually with a conspicuous hyaline awn at least on the older leaves; median leaf cells quadrate to rhombic or short-rectangular; cells in alar regions inflated in reddish groups, or not much differentiated; costa long-excurrent and forming the upper portion of the leaf, lamellose dorsally.

1. Campylopus pilifer Brid.
Figure 23, A–B

Campylopus introflexus (Hedw.) Brid. (in the sense of Breen, 1963).

Habitat: Dry, sandy soil in open pine forests.
Range: Europe; Africa; India; United States; Central America; South America.
Gulf Coast Distribution: Florida; Louisiana.

This species is easily recognized by the hyaline hair-points on the leaves. It is probably more widely distributed in our area than existing collections indicate.

Plants yellowish green, in thin low sods or dense tufts, or gregarious, sometimes producing deciduous, leafy buds in upper leaf axils; leaves subulate to narrowly lanceolate, not or only shortly decurrent, entire, or ± toothed at extreme tips, mostly 2–5 mm long, loosely appressed to spreading-ascending; median and upper

2. Campylopus tallulensis
Sull. & Lesq. ex Sull.
Figure 23, C–D

Campylopus flexuosus (in the sense of Breen, 1963).

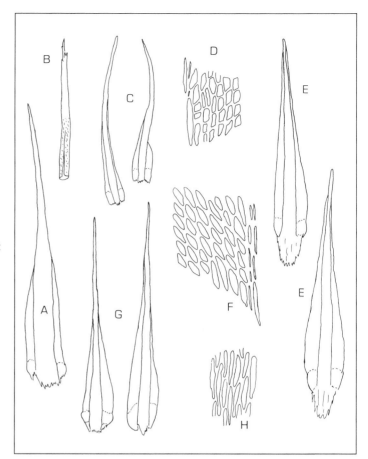

Fig. 23. *A–B:* **Campylopus pilifer.** *A,* leaf (× 15). *B,* hyaline leaf tip (× 55). *C–D:* **Campylopus tallulensis.** *C,* leaves (× 15). *D,* cells from broadest part of leaf (× 231). *E–F:* **Campylopus surinamensis.** *E,* leaves (× 15). *F,* cells from broadest part of leaf (× 231). *G–H:* **Campylopus angustiretis.** *G,* leaves (× 15). *H,* cells from broadest part of leaf (× 231).

cells irregularly subquadrate to short-rectangular or rhombic, alar cells mostly not forming conspicuous groups; costa percurrent to long excurrent and forming most of the upper portion of the leaf, smooth dorsally above.

Habitat: Sandy soil and sandstone in pine areas, including flatwoods.
Range: Eastern United States.
Gulf Coast Distribution: Florida to western Louisiana.

This *Campylopus* is fairly common in our area in longleaf pine habitats. Typically the plants have ± elongate stems, but in pine flatwoods a low fuzzy-appearing form occurs in which the stems are very short and the plants bear abundant, small, deciduous plantlets in their upper portions. Such plants have very short leaves and are quite different in appearance from the more typical form. The only other *Campylopus* in our area with which *C. tallulensis* could be easily confused is *C. surinamensis*; see comments under that species.

Similar in many respects to *C. tallulensis*; plants gregarious, often with a conspicuous terminal coma of larger leaves, leaves on stalk-like lower portion of the plant smaller and appressed; leaves mostly long-decurrent, lanceolate, mostly conspicuously roughened-toothed from the apex well down along the margins; median and upper cells mostly rhombic and in rows diverging outward from the costa; costa roughened dorsally in upper part.

Habitat: Sandy soil in ± open sites.
Range: Southeastern United States; tropical America.
Gulf Coast Distribution: Florida.

I have not seen any convincing material of this moss from the Gulf Coast west of Florida. The westernmost example I have seen was collected at Rock Hill, Washington County, Florida, near Chipley. However, this species could well occur further to the west. The capitate stems, rhombic leaf cells in diverging rows, and roughened to toothed leaf tips distinguish it from *C. tallulensis*. Also, *C. surinamensis* generally has a broader area of lamina in the middle and upper portions of the leaf than does *C. tallulensis* and mostly lacks a long-excurrent costa.

Plants in loose, soft tufts; lower leaves spreading, upper ones ± imbricate, mostly about 4 mm long, acuminate-subulate from lanceolate base; median cells linear-flexuous, thick-walled, alar cells inflated, reddish, forming conspicuous auricles; costa excurrent into a toothed awn.

Habitat: Sandy soil, cypress logs and stumps.
Range: Southeastern United States; Haiti; Puerto Rico.
Gulf Coast Distribution: Mississippi; southeastern Louisiana.

The elongate, thick-walled cells and conspicuously inflated alar cells forming auricles are distinctive. Although not yet reported from Alabama or the Florida Panhandle, this species probably occurs in both areas.

CAMPYLOPUS CAROLINAE Grout, described from the coastal plain of North Carolina, could occur in our area. The two stereid bands in its costa will distinguish it from all of our species of *Campylopus*.

Plants small to large, in dense clumps to low sods, mostly matted with tomentum below; stems ± erect, mostly simple; leaves mostly lanceolate-acuminate and ± secund, straight, curved, or sometimes ± contorted when dry; cells smooth or papillose, elongate, thick-walled, and porose throughout the leaf, or short and unpitted above and elongate and ± porose below, alar cells ± conspicuously differentiated; costa ending in the apex to shortly excurrent. Setae elongate; capsules ± asymmetric, straight and suberect to curved and inclined, ± sulcate when dry and empty; peristome single, the

3. **Campylopus surinamensis** C.M.
Figure 23, E–F
Campylopus gracilicaulis Mitt.
Campylopus donnellii Aust.

4. **Campylopus angustiretis** (Aust.) Lesq. & James
Figure 23, G–H

5. **DICRANUM Hedw.**

teeth 16, split to ⅔ down into 2 (or 3) divisions; operculum long-rostrate; calyptra cucullate.

Three species of *Dicranum* occur along the Gulf Coast; none of them is very common or abundant.

1. Upper cells elongate and porose; leaves usually distinctly secund   1. D. SCOPARIUM
1. Upper cells short, not or only slightly porose; leaves not usually distinctly secund   2
    2. Plants small, mostly growing on rotted pine logs and stumps; leaves mostly 2–3 mm long; slender, erect, flagelliform branches bearing reduced leaves commonly present in upper leaf axils   2. D. FLAGELLARE
    2. Plants medium-size, mostly growing on sandy soil and sandstone; leaves mostly about 5 mm long; flagelliform branches lacking   3. D. CONDENSATUM

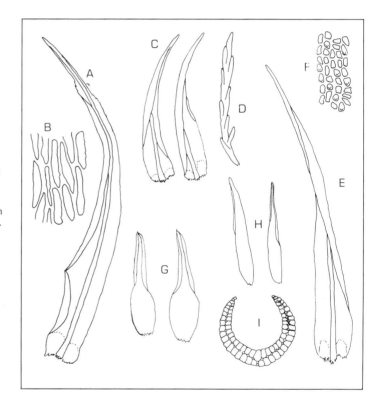

Fig. 24. *A–B:* **Dicranum scoparium.** *A,* leaf (× 15). *B,* cells from upper part of leaf (× 231). *C–D:* **Dicranum flagellare.** *C,* leaves (× 15). *D,* flagelliform branchlet (× 15). *E–F:* **Dicranum condensatum.** *E,* leaf (× 15). *F,* cells from upper part of leaf (× 231). *G,* **Leucobryum albidum,** leaves (× 11). *H–I:* **Leucobryum glaucum.** *H,* leaves (× 3). *I,* cross section of a leaf (× 55).

1. **Dicranum scoparium** Hedw.
Figure 24, A–B

Plants glossy, yellowish green to brownish, in loose tufts; leaves secund, curved, mostly 5–6 mm long, usually long and slenderly acuminate, not undulate, channeled above, apex toothed; margins serrate above; cells smooth, elongate and porose throughout, alar cells conspicuously enlarged, yellowish brown; costa ending in apex to shortly excurrent. Not producing sporophytes in our area.

Habitat: Rotted pine logs and stumps, rarely on humus; mixed forests.
Range: Widespread in the Northern Hemisphere.
Gulf Coast Distribution: Alabama, Louisiana, eastern Texas.

Dicranum scoparium is not a common moss on the Gulf Coast and apparently has not been reported from the Florida Panhandle or from coastal Mississippi; it likely occurs in both areas, however. The elongate, porose cells in the upper part of the leaf and the usually distinctly secund, nonundulate leaves distinguish *D. scoparium* from our other large *Dicranum*, *D. condensatum*.

Plants small, in low, dense turves or loose clumps, dull, dark green to brownish, slender, deciduous, flagelliform branches nearly always present (although sometimes scarce); leaves straight or curved, mostly 2–3 mm long, lanceolate, rather shortly acuminate, not undulate, channeled above; margins mostly entire; median and upper cells subquadrate, smooth, lower cells elongate, alar cells in distinct groups, hyaline to brownish; costa percurrent. Not producing sporophytes in our area.

2. Dicranum flagellare Hedw. Figure 24, C–D

Habitat: Rotted pine logs and stumps; bark of large tupelo and bald cypress trees in swamps; rarely on soil at tree bases.
Range: Widespread in the Northern Hemisphere.
Gulf Coast Distribution: Louisiana.

Although *D. flagellare* has apparently not been recorded from outside of Louisiana on the Gulf Coast, it very likely occurs all along the central and eastern portions of our area in suitable habitats. The plants are rather inconspicuous. Very small forms growing in low, dense turves, with leaves only 1–2 mm long, may be called var. *minutissimum* Grout. The flagelliform branches are almost uniformly present in specimens of *D. flagellare* and are distinctive.

Plants glossy, yellowish green to brownish, in loose tufts or turves; leaves straight or a little curved, mostly about 5 mm long, mostly not secund, broadly to slenderly acuminate, often undulate, keeled above, apex toothed, margins entire to serrate above; median and upper cells quadrate to short-rectangular, smooth to slightly papillose, thick-walled, not or only slightly porose, lower cells elongate and porose, alar cells hyaline to yellowish; costa mostly short-excurrent, often yellow at base. Infrequently producing sporophytes in our area; capsules curved, 1.5–2 mm long.

3. Dicranum condensatum Hedw. Figure 24, E–F

Habitat: Sandy soil, sandstone, rotted pine logs; rather dry, open forests inland, and old dunes along the coast.
Range: Eastern North America.
Gulf Coast Distribution: Florida to eastern Texas.

Dicranum condensatum is fairly frequent in dry, sandy areas along the Gulf Coast; the short upper cells and often undulate leaves easily distinguish it from *D. scoparium*. *Dicranum flagellare* is a much smaller moss that almost always bears slender, flagelli-form branches. The name *Dicranum sabuletorum* Ren. & Card. has been used for *D. condensatum* in some publications.

6. LEUCOBRYUM Hampe

Plants medium-size to large, in dense or loose turves or rounded cushions, mainly whitish; stems erect, simple or forked; leaves ± lanceolate, erect-appressed to spreading, thick, in cross section showing a single row of small green cells with one or more layers of large, empty cells above and below; costa very large, comprising most of the leaf above the base (the leaf is mostly costa). Setae elongate; capsules inclined, curved, ribbed when dry; peristome single, of 16 teeth split to the middle; operculum long and slenderly rostrate; calyptra cucullate.

Leucobryum is easy to recognize by the pale, whitish color (sometimes green), usually dense, soft colonies, and thick leaves. Two (or possibly three) species occur along the Gulf Coast. *Octoblepharum albidum* is similar but has flat, ligulate, spreading leaves with two layers of green cells as seen in cross section, and its peristome teeth are low, blunt, and undivided.

Leucobryum is often treated separately in its own family, the Leucobryaceae, due to its very characteristic gametophytes. Its sporophyte, however, clearly shows its relationship with the Dicranaceae.

Plants small, often fertile, in dense, compact colonies; leaves straight, mostly less than 4 mm long, base mostly about as long as the upper part	1. L. ALBIDUM
Plants medium-size to large, not fruiting in our area, in loose or compact colonies; leaves often spreading-flexuous, mostly longer than 4 mm, base shorter than the upper part	2. L. GLAUCUM

1. Leucobryum albidum (Brid.) Lindb.
Figure 24, G

Small plants in usually dense, tight sods or cushions; leaves mostly ascending or erect, short, mostly about 3 mm long, not flexuous, the base about as long as the upper part. Rather commonly producing sporophytes; capsules about 1 mm long.

Habitat: Rotted logs, stumps, humus, sandy soil, rocks; mesic to dry forests.
Range: Europe; northern Africa; eastern United States; Mexico; Caribbean.
Gulf Coast Distribution: Florida to eastern Texas.

This species is fairly common over most of our area. Fertile specimens are most often found on rotted pine logs and stumps. The compact growth habit, small size, and short leaves generally distinguish it easily from the much scarcer *L. glaucum*.

Medium-size to robust plants in loose to rather compact mats or cushions; leaves erect to ± spreading, straight to flexuous, 3–8 mm long, the base usually shorter than the upper part. Apparently not producing sporophytes in our area; capsules 1.5–2 mm long.

Habitat: Sandy soil, humus, rotted logs; ravines, stream banks, tree bases; wet to mesic forests.
Range: Europe; Japan; northern Africa; eastern North America.
Gulf Coast Distribution: Florida to eastern Texas.

This species, sometimes called "pincushion moss" because of the rounded colonies it often forms, is rather uncommon along the Gulf Coast. It occurs in generally moister and more sheltered habitats than does *L. albidum*. I have not seen any fertile specimens of *L. glaucum* from the Gulf Coast. See comments under *L. albidum*.

LEUCOBRYUM ANTILLARUM Schimp. ex Besch. has been reported from several areas along the Gulf Coast, but the specimens so identified may really be robust forms of *L. glaucum* with longer and more flexuous leaves than is usual for that species. The published distinctions between *L. antillarum* and *L. glaucum* are not clear-cut.

2. **Leucobryum glaucum**
(Hedw.) Ångstr. ex Fries
Figure 24, H–I

CALYMPERACEAE Kindb.

Plants small to medium-size, mostly dark green, sometimes yellowish green or paler, gregarious or in compact sods or tufts; stems mostly erect, often forked; leaves linear to lanceolate or ligulate, mostly appressed when dry, sometimes spreading, with or without a border of elongate, hyaline cells, often bearing septate gemmae, rarely several cell layers thick and then consisting mostly of the costa; upper cells ± isodiametric, smooth to papillose, basal cells enlarged, porose, hyaline, forming characteristic areas on both sides of the costa and termed cancellinae; marginal or intramarginal bands of elongate, hyaline cells (teniolae) sometimes present (lacking in our taxa); costa single, strong. Sporophytes rarely produced, terminal; setae mostly elongate; capsules erect; peristome lacking or single and of 8 or 16 short, blunt teeth; operculum rostrate; calyptra cucullate (in our species), or clasping the seta below the capsule and persistent.

This is primarily a family of the tropics; two genera occur in our area. The cancellinae, and the hyaline leaf border on some of the species, are characteristic and combine to make these mosses very attractive under the microscope.

Plants green to brownish; leaves (except the costa) one cell layer thick, mostly lanceolate, mostly ascending and ± contorted when dry

1. SYRRHOPODON

Plants white to greenish white; leaves several cell layers thick, flat,
   ligulate, spreading when dry   2. OCTOBLEPHARUM

1. SYRRHOPODON
Schwaegr.

Plants small to medium-size, yellowish to dark green or brownish, in compact sods or small tufts, sometimes gregarious; leaves lanceolate to ligulate, basal part, including the cancellinae, usually distinctively differentiated from the green upper part; margins mostly either thickened and serrate or bordered with elongate hyaline cells and variously toothed; cells of upper part of leaf mostly ± quadrate, smooth to papillose; costa strong. Setae mostly elongate; capsules cylindric; peristome lacking or of 16 short, often imperfect teeth; calyptra cucullate.

Five species of this mainly tropical genus occur along the Gulf Coast. Sporophytes are only rarely produced by some of our species of *Syrrhopodon*, and not at all by others. Reproduction and dispersal are effected instead by the gemmae, which are often formed in great abundance on the leaf tips or, in the case of *S. parasiticus*, along the costa on the upper side of the leaf. The septate gemmae dehisce readily and are dispersed by air currents. The characteristic empty cells in the leaf bases, comprising the cancellinae, are porose and function as storage sites for water. Rhizoids often grow into the cells of the cancellinae, perhaps serving to transfer water from the cancellinae to other parts of the plant.

Four of our five species of *Syrrhopodon* have the greater part of their ranges to the south of the Gulf Coast—in the American tropics. It is interesting to speculate on how such plants arrived on our shores. Did they migrate slowly from island to island across the Caribbean? Were they carried as spores or fragments on hurricane winds from the south? Were they part of an ancient tropical flora that once occupied North America long ago, when the climate was warmer, and then were left behind as the climate cooled and the tropical flora retreated south? Perhaps each of these hypotheses (or combinations of them) is true for individual cases. The fifth species of *Syrrhopodon*, *S. texanus*, is a native North American species that does not occur in the tropics.

1. Leaf margins bordered entirely or in part with elongate, hyaline cells   2
1. Leaf margins thickened but lacking a border of elongate cells   1. S. INCOMPLETUS
   2. Lower portion of leaf strongly toothed on the margins   2. S. TEXANUS
   2. Lower portion of leaf with entire margins or occasionally with
      small, irregular teeth   3
3. Leaves linear, bordered all around with elongate cells; plants
   terrestrial   3. S. PROLIFER
3. Leaves ligulate to lanceolate, border of elongate cells often present only on lower portion of leaf; plants epiphytic   4

4. Plants minute; leaves tightly crispate when dry; gemmae
    borne on leaf tips   4. S. LIGULATUS

4. Plants to 1 cm tall; leaves variously curved or twisted when
    dry but not crispate; gemmae borne along costa on upper
    surface of leaf   5. S. PARASITICUS

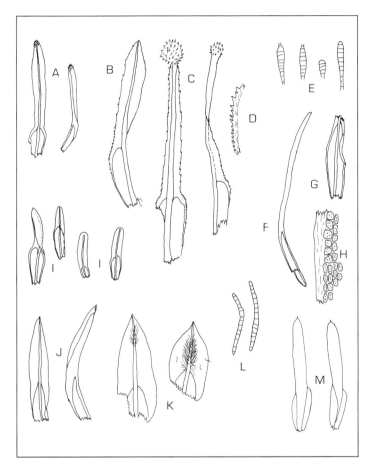

Fig. 25. *A,* **Syrrhopodon incompletus,** leaves (× 9). *B–E:* **Syrrhopodon texanus.** *B,* leaf (× 15). *C,* modified gemmiferous leaves (× 15). *D,* margin at leaf shoulder (× 55). *E,* gemmae (× 55). *F–H:* **Syrrhopodon prolifer.** *F,* leaf (× 15). *G,* leaf base (× 15). *H,* margin and cells from upper part of leaf (× 231). *I,* **Syrrhopodon ligulatus,** leaves (× 15). *J–L:* **Syrrhopodon parasiticus.** *J,* leaves (× 9). *K,* enlarged comal leaves bearing gemmae (× 9). *L,* gemmae (× 55). *M,* **Octoblepharum albidum,** leaves (× 8).

Plants mostly dark green to brownish, in thin to dense colonies; stems to about 1 cm tall but mostly shorter in our area; leaves lanceolate, mostly 4–5 mm long; margins of upper part thickened and coarsely toothed in two rows; upper leaves frequently bearing gemmae on their tips. Not often fruiting in our area.

1. **Syrrhopodon incompletus**
Schwaegr.
Figure 25, A

Habitat: Tree bases, especially on live oak, bald cypress, and
    palmetto, in forests; occasionally on sandstone. Often grow-
    ing with other mosses, especially *Sematophyllum adnatum*
    and *Fissidens cristatus.*
Range: Eastern United States; tropical America.
Gulf Coast Distribution: Florida to southwestern Louisiana.

This is a mainly tropical species whose range in North America is confined to the coastal plain. It has been found as far north as Long Island, New York, but its main distribution in the United States is the Gulf coastal plain, never far from the Gulf of Mexico. It is very common and abundant in Florida, and will likely be found eventually in southeastern Texas. In older works on mosses this species is usually referred to as *S. floridanus* Sull. Other forms of *S. incompletus* are known in the southern part of its range; ours is most precisely named *S. incompletus* var. *incompletus*. (The name *incompletus* refers to the lack of a peristome in the capsules of many specimens.) The absence of elongate cells on the leaf margins distinguishes *S. incompletus* from all of our other species of the genus.

2. **Syrrhopodon texanus** Sull.
Figure 25, B–E

Dull, dark green plants growing in dense cushions which often appear bristly due to the stiffly erect gemmiferous leaves; stems mostly about 1 cm tall; leaves usually of two types, one broad and vegetative, the other very narrow and bearing gemmae on the tip, both types bordered all around with elongate hyaline cells and coarsely toothed in the lower portion on the margins; cells obscure, densely papillose. Sporophytes infrequently produced.

> **Habitat:** Tree bases, stumps, rotting wood, soil, rocks, in low forests, often in wet sites.
> **Range:** Eastern North America, northeastern Mexico.
> **Gulf Coast Distribution:** Florida to central Texas.

Syrrhopodon texanus is the most common and abundant of the five species of the genus that are known along the Gulf Coast. It is also the only member of the genus that has an exclusively North American distribution and that occurs in the interior of the continent as well as on the coastal plain. The stoutly toothed margins on the lower portion of the leaf distinguish *S. texanus* from our other species of the genus.

3. **Syrrhopodon prolifer**
Schwaegr.
Figure 25, F–H

Yellowish green, pale plants growing in grasslike clumps or thin colonies; stems to about 1 cm tall; leaves linear, straight, bordered all around with elongate cells, margins mostly entire except toothed at apex; cells densely papillose; gemmae scanty. Not forming sporophytes in our area.

> **Habitat:** On sandy soil, at tree bases in the bottoms of shaded, humid ravines.
> **Range:** Southeastern U.S.; tropical America.
> **Gulf Coast Distribution:** Panhandle Florida, southeastern Louisiana.

This species is rare on the Gulf Coast although very common and often abundant in the American tropics. In our area it is known only from a few sites in the Panhandle counties of Florida and from two stations close together in Washington Parish, Louisiana. Several taxonomic varieties of this species are known. The one which occurs along the Gulf Coast is *S. prolifer* Schwaegr. var. *papillosus* (C.M.) Reese.

Plants of *S. prolifer* are very attractive under the microscope. The hyaline margins and prominent cancellinae give an air of delicate beauty to the leaves at low magnification. Additional stations for this moss should turn up following a search for it in appropriate habitats between Florida and Louisiana. Deep, humid, forested ravines with springy seeps along the bottoms are prime places to look.

Very small plants with tightly crispate leaves when dry, growing in dense sods or, more often in our area, as scattered individuals or clumps of a few stems; leaves ligulate, about 2 mm long, border of elongate cells usually incomplete, often confined to the leaf base or extending just above the cancellinae; cells very densely papillose; leaf tip often notched; a few teeth sometimes present on leaf margin at midleaf. Not forming sporophytes in our area.

4. Syrrhopodon ligulatus
Mont.
Figure 25, i

Habitat: Tree trunks, especially bald cypress, in swamps and low forests, occasionally on rock.
Range: Southeastern U.S., tropical America.
Gulf Coast Distribution: Florida, Georgia, Mississippi, southeastern Louisiana.

The small stature and inconspicuous habit of *S. ligulatus* make it difficult to find in the field. It is probably much more common than existing collections along the Gulf Coast indicate. The short, ligulate leaves with the apex often notched and the incomplete border of elongate cells make this species easy to identify.

Plants small to medium, dark green, in small clumps of a few stems; leaves involute and incurved when dry, spreading when moist, linear-lanceolate to broadly lanceolate, often triangular in rosettelike accretions at stem tips; upper leaves frequently bearing abundant filiform gemmae along the costa on the upper surface; cells smooth or with low papillae; leaf margins bordered entirely or partly with elongate cells. Not forming sporophytes in our area.

5. Syrrhopodon parasiticus
(Brid.) Besch.
Figure 25, J–L

Habitat: Bark of tree trunks, branches, and twigs, in humid forests.
Range: Southeastern U.S., tropical America.
Gulf Coast Distribution: Florida to eastern Texas.

Syrrhopodon parasiticus grows in such small colonies—and sometimes as single stems—that it is easy to miss it in the field. However, if one searches carefully in likely habitats it can usually be discovered. Likely habitats include wet forests, around springs in woods, along streams, and on river banks. This moss favors smooth-barked trees such as holly and maple, but·has been found on many other types of trees as well. Its abundantly produced gemmae apparently serve well for reproduction and dispersal because it is known from many collections along the Gulf Coast. It is fairly common but never abundant.

2. OCTOBLEPHARUM Hedw.

Octoblepharum albidum
Hedw.
Figure 25, M

Plants small, white to brownish or greenish white, sometimes tinged with pink, in dense, low turves or clumps; stems erect, mostly simple; leaves spreading when dry, flat, apiculate, ligulate from a broader, winged base, composed mostly of the costa, several cell layers thick, in cross section showing one or two layers of small green cells with several layers of large empty cells above and below, lamina consisting of hyaline wings restricted to the leaf base. Setae elongate; capsules erect, ovoid-cylindric, brown; peristome of 8 short, blunt teeth.

> **Habitat:** Bark and bases of trees, especially live oak and palm.
> **Range:** Pantropical.
> **Gulf Coast Distribution:** Florida; Louisiana; Texas.

Octoblepharum albidum is so distinctive in its flat, ligulate leaves that it is very easy to recognize. It is rare in our area, where it is known from a few collections in New Orleans and one, made long ago, in the vicinity of Matagorda Bay, Texas. It should be sought on large old live oaks and palmettos in sheltered places along the coast.

Octoblepharum has traditionally been placed with *Leucobryum*, in the family Leucobryaceae. However, its peristome indicates a closer alliance with the Calymperaceae although there are difficulties in explaining the gametophytic relationships. The genus *Leucobryum* itself is best considered as a member of the Dicranaceae.

POTTIACEAE Schimp.

Plants very small to medium-size or robust, yellowish to dark green or brownish, dull, mostly growing in low, compact sods or tufts; stems erect, often forked, usually very short; leaves mostly variously contorted when dry, linear to broadly or narrowly lanceolate, mostly acute; margins plane to revolute or involute, mostly entire; upper cells mostly small, firm, isodiametric, usually strongly papillose and ± obscure, lower cells larger, often ± hyaline and smooth, not differentiated in alar regions; costa single, strong, mostly percurrent to short-excurrent, in cross section showing either dorsal and ventral bands of stereid cells separated by a median row of large, empty guide cells, or only a dorsal band of stereids. Setae very short to elongate; capsules erect and symmetric, sometimes

Table 2. Selected Characteristics of Gulf Coast Genera of Pottiaceae

	Leaf margins conspicuously infolded-involute (I) or recurved-revolute (R)	Cleisto-carpous; capsules ± immersed	V-shaped area of hyaline cells in leaf base	Brood bodies commonly present	Eastern (E) or western (W) distribution on Gulf Coast	Corticolous	± Strict calciphile	Leaf apex awned	Always sterile on Gulf Coast
Acaulon		*							
Aloina	I						*		*
Astomum	I	*							
Barbula	R, pp		pp				pp		
Crossidium	R				W			*	*
Desmatodon	R						*	*	
Eucladium					E		*		*
Gymnostomiella			*		E		*		*
Gymnostomum							*		
Husnotiella	R				W				
Hyophila				*					
Luisierella			*		E		*		
Phascum		*			W			*	
Pterygoneurum					W			*	
Splachnobryum							*		*
Tortella			*				pp		
Tortula				pp		pp		pp	*
Trichostomum							*		*
Tuerckheimia					E		*		*
Weissia	I		pp						

pp = in part

immersed and cleistocarpous; peristome single, the 16 teeth often spirally twisted, sometimes split into 32 slender divisions, or lacking; operculum conic to rostrate, sometimes not differentiated; calyptra cucullate.

This is a large and complex family of mainly terrestrial mosses; many of the species are calciphiles and a few grow on the bark of trees. As a general rule, the combination of low, dull plants with contorted leaves and small, highly papillose cells, and the mainly terrestrial habitat, is indicative of the Pottiaceae.

Table 2 indicates some of the outstanding or otherwise taxonomically useful features of Gulf Coast Pottiaceae, by genus, and may be used as a preliminary guide to identification. For example, if you have a specimen that has conspicuously recurved margins and an awned leaf tip and that was growing on a limey substrate, it is quite likely to turn out to be *Desmatodon plinthobius*. However, the table should be used only as a tentative guide to identification; always check the description and illustrations too.

1. Costa bearing ventral lamellae or filaments; plants known from Louisiana and Texas in our area   2
1. Costa lacking ventral appendages; distribution various   4

2. Costa densely covered with filaments 3
2. Costa bearing several ventral lamellae 13. Pterygoneurum
3. Leaves with prominent hair-points, margins revolute 14. Crossidium
3. Leaves lacking hair-points, margins infolded 16. Aloina
  4. Leaf margins involute, wet and dry 5
  4. Leaf margins plane to recurved or revolute, at least when moist (slightly incurved at apex in *Tortella flavovirens*; incurved when dry in *Hyophila*) 7
5. Cleistocarpous; capsules immersed to barely exserted; plants not restricted to calcareous substrates 1. Astomum
5. Stegocarpous (one always sterile in our area, and a calciphile); capsules well exserted; substrates various 6
  6. Plants sterile in our area, restricted to limey substrates; basal hyaline cells extending up along margins and often forming a V-shaped area 2. Weissia jamaicense
  6. Plants usually or often fertile, substrate various; basal cells not hyaline or not forming V-shaped border 2. Weissia controversa
7. Plants small, bulbiform; upper leaves broadly ovate; capsule immersed 11. Acaulon
7. Plants small to large, not bulbiform; upper leaves narrow or broad; capsules immersed to exserted 8
  8. Costa with only a dorsal band of stereid cells, or cells of costa ± homogeneous (as seen in cross section) 9
  8. Costa with both dorsal and ventral bands of stereid cells (as seen in cross section) 16
9. Costa ending below leaf apex; leaf apex broadly rounded 10
9. Costa excurrent; leaf apex mostly acute 13
  10. Leaf cells smooth or bulging, not papillose 11
  10. Leaf cells papillose 12
11. Leaf cells smooth; leaves lingulate-spatulate, lacking V-shaped area of hyaline cells in base 19. Splachnobryum
11. Leaf cells bulging; leaves ligulate, with V-shaped area of hyaline cells in base 20. Luisierella
  12. Leaf cells densely papillose, obscure; leaf margins recurved; costa often spurred above 4. Husnotiella
  12. Leaf cells lightly papillose, pellucid; leaf margins plane; costa not spurred 18. Gymnostomiella
13. Plants mainly corticolous; leaf margins plane; small, rough, leaflike propagula common in upper leaf axils 17. Tortula pagorum
13. Plants terrestrial; leaf margins plane to ± recurved at least in part; axillary propagula lacking 14
  14. Leaf margins plane; cells smooth to delicately papillose 17. Tortula rhizophylla
  14. Leaf margins recurved, at least in part; cells mostly strongly papillose 15
15. Leaf margins recurved at midleaf only; capsules immersed, cleistocarpous 12. Phascum
15. Leaf margins usually recurved throughout; capsules exserted, with operculum and peristome 15. Desmatodon
  16. Hyaline basal cells of leaf extending up margins and forming a V-shaped area 8. Tortella
  16. Hyaline basal cells various or lacking but not forming a V-shaped area 17

17. Leaf apex usually obtuse, mucronate, and irregularly
    coarsely toothed; stalked, echinate gemmae common
    in upper leaf axils   9. Hyophila
17. Leaf apex obtuse to acute, not toothed; gemmae, if present, not
    echinate   18
    18. Lower leaf margins toothed on at least some leaves; basal
        cells conspicuously enlarged and hyaline   5. Eucladium
    18. Lower leaf margins essentially entire (except for papillae
        sometimes); basal cells not conspicuously enlarged or
        hyaline   19
19. Leaf margins plane throughout   20
19. Leaf margins recurved, at least below   23
    20. Upper leaf cells bulging ventrally, flat dorsally, not
        papillose   10. Barbula agraria
    20. Upper leaf cells papillose   21
21. Plants small, ± glossy when dry; leaves mostly less than 3 mm
    long, tips not fragile and broken off   22
21. Plants medium-size, dull when dry; leaves more than 3 mm
    long, tips fragile and often broken off   6. Tuerckheimia
    22. Largest leaves to about 1.5 mm long; costa ending below
        leaf apex   3. Gymnostomum
    22. Longest leaves to about 2.5 mm long; costa excurrent into
        a stout point   7. Trichostomum molariforme
23. Plants small, ± glossy, calciphile; costa ending below leaf tip   3. Gymnostomum
23. Plants medium-size, dull, habitat various; costa percurrent to
    short-excurrent   10. Barbula

1. ASTOMUM Hampe

Plants small, in dense, dark green clumps or sods; stems mostly short, simple or forked, erect; leaves mostly strongly contorted when dry, erect-spreading when moist, lanceolate to linear-lanceolate, acute; margins mostly strongly involute both wet and dry; upper cells small, papillose, obscure; lower cells somewhat larger, smooth, pale; costa strong, percurrent to slightly excurrent. Setae very short; capsules mostly immersed, ± spherical to elongate, lacking peristome and operculum, the latter sometimes ± distinct but not dehiscing, with short to long apiculation.

Species of *Astomum* are often treated in the closely related genus *Weissia*. The only basic difference between *Astomum* and *Weissia* is the cleistocarpous capsule of the former. When sterile plants only are at hand, it may be difficult to decide whether they belong to *Astomum* or *Weissia*. However, comparison with previously determined, fertile material will usually permit identification. Plants of *Astomum* grow on soil in damp, ± disturbed sites in weedy or forested areas. Three species of *Astomum* occur along the Gulf Coast; two of them are fairly common: *A. ludovicianum* and *A. muhlenbergianum*. Both of these species are known to hybridize with *Weissia controversa*, an indication of their close relationship to *Weissia*.

1. Seta very short; capsule appearing to be sessile, ± spherical, with short apiculation .... 2. A. MUHLENBERGIANUM
1. Seta evident; capsule not appearing sessile, spherical to elongate, with short or long apiculation. 2
    2. Capsule elongate, with short, blunt apiculation; seta usually shorter than the capsule .... 1. A. LUDOVICIANUM
    2. Capsule ± spherical to somewhat elongate, with long, slender, oblique rostrum; seta usually as long as or longer than the capsule .... 3. A. PHASCOIDES

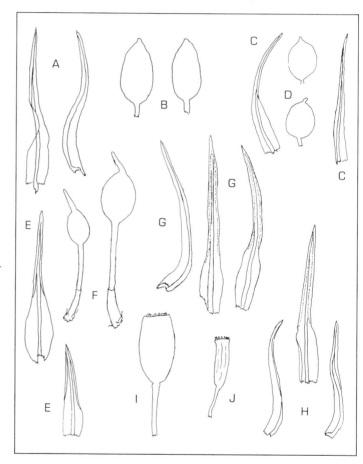

Fig. 26. *A–B*: **Astomum ludovicianum.** *A*, leaves (× 15). *B*, capsules (× 15). *C–D*: **Astomum muhlenbergianum.** *C*, leaves (× 15). *D*, capsules (× 15). *E–F*: **Astomum phascoides.** *E*, leaves (× 15). *F*, sporophytes (× 15). *G*, **Weissia jamaicensis**, leaves (× 15). *H–J*: **Weissia controversa.** *H*, leaves (× 15). *I*, soaked capsule (× 15). *J*, dry capsule (× 15).

1. **Astomum ludovicianum** (Sull.) Sull.
Figure 26, A–B

Plants small to medium-size, tufted or in sods; upper leaves very slender, to about 5 mm long; margins occasionally plane in part. Commonly fertile; capsules elongate, broadest at base, glossy, with short, blunt apiculation, operculum often set off by a ± distinct line; sporophytes often 2–3 per perichaetium.

    Habitat: Wet to moist, often clayey soil; weedy sites and in forests; ditch banks and roadsides.

Range: Eastern United States.
Gulf Coast Distribution: Florida to eastern Texas.

This moss has apparently not yet been reported from Mississippi but surely occurs there. *Astomum ludovicianum* can be found with sporophytes from November through July in our area but is most commonly found fertile from January to March. It is easily identified by its elongate capsules with short, blunt apiculus, and evident seta.

Plants smaller than those of *A. ludovicianum*; upper leaves to 3–4 mm long. Setae very short and inconspicuous, the capsules appearing to be sessile; capsules ± spherical, usually concealed within the perichaetial leaves when dry, with short, blunt apiculus; the operculum not set off by a line.

> **Habitat:** Soil in weedy or brushy sites; roadsides and ditch banks; old fields; usually in drier areas than *A. ludovicianum*.
> **Range:** Asia; eastern North America.
> **Gulf Coast Distribution:** Mississippi to Texas.

This species has not yet been reported from Panhandle Florida or from Alabama, but doubtless occurs in both areas. The gametophytes are very similar, usually, to those of *Weissia controversa*. The essentially sessile, ± spherical capsules, with short apiculus, make *A. muhlenbergianum* easy to recognize. It is not as common as *A. ludovicianum* and produces its sporophytes during the same period.

Generally similar to *A. ludovicianum* but differing in the usually more spherical capsules, the long, slender, oblique rostrum, and the longer setae.

> **Habitat:** Soil on banks, roadsides; open or forested areas; old fields.
> **Range:** Europe; North America.
> **Gulf Coast Distribution:** Texas.

Astomum phascoides is rare in our area, where it has been reported from a few counties in Texas. The relatively long seta and slender, oblique rostrum make it easy to identify.

Small to medium-size plants in green to yellowish or brownish green tufts or sods; stems short, erect, simple or forked; rhizoids sometimes bearing tubers; leaves largest toward stem tips, variously contorted and curled when dry, erect-spreading when wet, oblong to lanceolate or linear-lanceolate, pointed at apex, base pale; margins entire, involute above both wet and dry, and upper part of leaf thus channeled; upper cells small, obscure, densely papillose; costa strong, shortly excurrent into a pale point. Setae terminal,

2. **Astomum muhlenbergianum** (Sw.) Grout
Figure 26, C–D

3. **Astomum phascoides** (Hook. ex Drumm.) Grout
Figure 26, E–F

2. **WEISSIA Hedw.**

elongate, yellowish; capsules erect, straight or slightly curved and asymmetric; peristome lacking or present and of 16 rudimentary or well-developed straight, papillose teeth, the teeth entire or ± divided or perforated; spores finely papillose; operculum obliquely rostrate; calyptra cucullate, smooth.

There are at least two species of *Weissia* in our area; *W. controversa* is weedy, widespread, and common, sometimes conspicuous in the early spring by its abundance and yellow setae. Another species, *W. jamaicense*, is less common and more restricted in its habitat.

Plants ± dull, coarse, often appearing glaucous, calciphile; leaves mostly 3–4 mm long; hyaline basal cells often forming a V-shaped area	1. W. JAMAICENSE
Plants ± glossy, not appearing glaucous, not restricted to calcareous habitats; leaves mostly 1–2 mm long; V-shaped area lacking in base	2. W. CONTROVERSA

1. **Weissia jamaicense** (Mitt.) Grout
Figure 26, G

Trichostomum jamaicense (Mitt.) Jaeg. & Sauerb.

Plants medium-size, glaucous-green to brownish, in low, dull, wirey clumps or tufts (the hyaline, glossy leaf bases sometimes conspicuous); stems erect, mostly simple; leaves strongly but loosely contorted when dry, the tips often curled, stiffly erect-spreading when moist, subulate from a broad, somewhat sheathing, ± hyaline base, mostly 3–4 mm long; upper cells usually very obscure, densely pluripapillose (papillosity sometimes difficult to see because the papillae may be very dense), lower cells smooth, rectangular, hyaline, the hyaline cells usually extending up the margins somewhat as in *Tortella*; margins strongly involute above the leaf base, sometimes only in the upper ¼–⅓ of the leaf; costa very stout, mostly 60 μm or more wide at the leaf shoulders, mostly shortly excurrent as a pellucid mucro. Apparently dioicous; sterile in our area, but archegoniate plants are frequent.

> **Habitat:** Limestone, calcareous soil, old concrete, rarely on rotted wood; ± open sites, including road-cut banks.
> **Range:** Southern and southwestern United States; Caribbean; Mexico; Guatemala.
> **Gulf Coast Distribution:** Florida; Alabama; Louisiana; Texas.

This species is locally common on limestone and calcareous soils in Panhandle Florida and Louisiana. It probably occurs throughout our area. The plants could be taken for a *Tortella* because of the hyaline basal cells that usually extend up the leaf margins, but the leaves in *Tortella* mostly have the margins plane or at most slightly incurved and then only at the apex. *Weissia jamaicense* is more likely to be confused with the common and weedy *W. controversa*, but that species has smaller plants with shorter, more tightly contorted leaves, and a narrower costa. Also, plants of *W. controversa* are generally ± glossy and are not restricted to calcareous substrates. *Weissia jamaicense* is easy to recognize, even in the field, by its dull, glaucous-green to brownish appearance, curled leaf tips, and calcareous habitat.

Plants small, in light to yellowish green, compact, low sods; rhizoids sometimes bearing pale purple tubers; stems short, erect, forked; leaves strongly curled when dry with the glossy dorsal surface of the costa conspicuous, mostly 2–3 mm long, erect-spreading when moist, narrowly- to linear-lanceolate, acute, margins strongly involute wet and dry; upper cells isodiametric, small, obscure, densely pluripapillose, basal cells larger, ± hyaline; costa strong, short-excurrent. Commonly producing sporophytes; setae yellow, longer to much longer than the perichaetial leaves; capsules erect, ± globose to cylindric, symmetric or sometimes curved, yellowish, becoming darker and furrowed with age; peristome teeth often short and irregular, often fallen from older capsules; spores finely papillose; operculum obliquely long-rostrate; calyptra cucullate.

2. **Weissia controversa** Hedw.
Figure 26, H–J

> **Habitat:** Disturbed sites on bare calcareous or sandy soil; rocks; sometimes on tree bases and rotted logs; roadsides, old fields, forests; virtually everywhere.
> **Range:** Cosmopolitan.
> **Gulf Coast Distribution:** Florida to Texas.

Weissia controversa is a common and weedy little moss throughout our area. The yellowish setae and tightly contorted leaves, when dry, are distinctive. *Weissia jamaicense* is similar but coarser, sterile in our area, is restricted to calcareous substrates, and has a V-shaped area of hyaline cells in the leaf base. Plants of *Astomum* may be quite similar to those of *Weissia controversa*; it may sometimes be difficult to identify sterile plants with certainty. However, once *W. controversa* has been learned it can be recognized easily both fertile and sterile. See comments under *Astomum*. *Ptychomitrium incurvum* might also be confused with *Weissia controversa*. Specimens of *W. controversa* with the seta elongate (to approximately 25 mm) are sometimes recognized as *W. controversa* var. *longiseta* (Lesq. & James) Crum, Steere and Anderson.

WEISSIA HEDWIGII Crum [as *W. microstoma* (Hedw.) C.M.] and W. ANDREWSII Bartr. have both been reported from our area in Texas but must be very scarce there, if actually present. The former differs from *W. controversa* in lacking a peristome altogether and in having the mouth of the capsule ± closed by a membrane (hymenium); the latter differs from *W. controversa* in having a wider costa (to *ca* 65 μm versus *ca* 50 μm in *W. controversa*) and ovate- to oblong-lanceolate leaves; its peristome is similar to that of *W. controversa*.

3. GYMNOSTOMUM Nees & Hornsch.

Gymnostomum aeruginosum Sm.
Figure 27, A

Gymnostomum calcareum Nees, Hornsch. & Sturm (in the sense of Breen, 1963)

Plants small, somewhat glossy, green to yellowish green at growing tips, brown below, in loose or dense turves or (sometimes deep) cushions; stems mostly simple; leaves ± contorted when dry, 1–2 mm long, spreading to recurved when moist, narrowly- to linear-lanceolate, acute to obtuse; margins plane, sometimes recurved below, often thickened above; upper cells subquadrate, pluripapillose, the papillae small to large, lower cells rectangular, smooth, mostly yellowish, grading into upper cells; costa mostly ending below leaf tip, sometimes percurrent. Sporophytes small; setae elongate; capsules ovoid to elliptic; peristome lacking; spores smooth to finely granular; operculum obliquely rostrate; calyptra cucullate; probably sterile in our area.

> **Habitat:** Moist calcareous rock; old bricks and mortar; occasionally on sandstone.
> **Range:** Europe; Asia; Africa; North, South, and Central America.
> **Gulf Coast Distribution:** Florida; Louisiana.

This little calciphile occurs naturally on native limestone in Florida but has also been collected several times on old tombs in cemeteries in New Orleans. The habitat, small size, plane margins, and costa usually ending well below the leaf tip are distinctive. The leaves can be blunt and rounded at the tips or acute, in a single collection. *Gymnostomum aeruginosum* could exist on limestone elsewhere along the Gulf Coast, especially in Texas.

4. HUSNOTIELLA Card.

Husnotiella revoluta Card.
Figure 27, B–C

Plants small, green to brownish, in low, dense sods; stems erect, simple; leaves somewhat contorted when dry, spreading when moist, lingulate to lingulate-lanceolate or ovate, concave, mostly bluntly rounded and often cucullate at apex, sometimes acute, 1–1.5 mm long; margins closely revolute above, plane below, entire; upper cells obscure, rounded-quadrate, mammillose to papillose, lower cells hyaline, rectangular, smooth; costa very stout, clearly ending several cells below the leaf tip, often spurred along the sides and at the tip. Setae elongate; capsules ovoid to cylindric; peristome lacking; spores smooth to finely granular; operculum obliquely rostrate; calyptra cucullate.

> **Habitat:** Calcareous soil.
> **Range:** Southwestern United States; Mexico.
> **Gulf Coast Distribution:** Southern Texas.

In its typical form, with rounded leaf tips and spurred costa, *H. revoluta* is easy to recognize. It is known in our area from Cameron County, Texas, and also occurs in the mountains of western Texas. In some specimens the upper surface of the costa is very rough near the leaf tip, with bulging cells, giving somewhat the impression of low filaments. I have not seen the specimen from Cameron County, Texas, and it is possible that this species does not actually occur in our area.

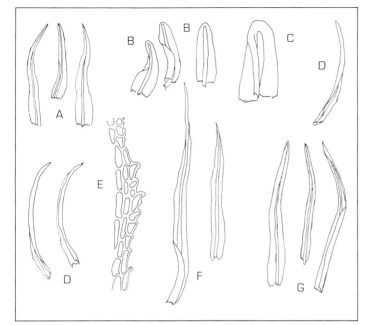

Fig. 27. *A,* **Gymnostomum aeruginosum,** leaves (× 23). *B–C:* **Husnotiella revoluta.** *B,* leaves (× 15). *C,* leaf tip (× 55). *D–E:* **Eucladium verticillatum.** *D,* leaves (× 15). *E,* margin at leaf shoulder (× 231). *F,* **Tuerckheimia angustifolia,** leaves (× 15). *G,* **Trichostomum molariforme,** leaves (× 15).

Plants small, yellowish green to green, in compact sods; stems erect, branched; leaves straight to somewhat contorted when dry, wide-spreading to recurved when moist, linear-lanceolate, narrowly acute, 2–3 mm long; margins plane, toothed (at least some leaves) at junction of green cells and hyaline basal cells; upper cells ± quadrate, with one to several usually prominent papillae on both surfaces, basal cells enlarged, hyaline, smooth, sharply distinct from the green upper cells; costa ending in leaf apex or shortly excurrent. Not producing sporophytes in our area.

> **Habitat:** Moist limestone.
> **Range:** Widespread in the Northern Hemisphere.
> **Gulf Coast Distribution:** Florida.

This attractive little moss is easy to identify because of its habitat, narrow leaves, marginal teeth at the leaf shoulders, and enlarged, hyaline basal cells. It is quite likely that *E. verticillatum* will turn up in calcareous habitats in coastal areas of Texas.

Plants small, glaucous to green or brownish green, in dense turves; stems erect, simple, often with red rhizoids in lower parts; leaves contorted when dry, spreading when moist, 3–5 mm long, linear-lanceolate, narrowly acute at apex, fragile and often with the tips broken off; margins plane, sometimes sinuous near the apex; upper cells thick-walled, quadrate to transversely elliptic, with large, blunt papillae on both surfaces, lower cells larger, smooth, rectangular; costa mostly shortly excurrent. Not producing sporophytes in our area.

5. EUCLADIUM B.S.G.

Eucladium verticillatum (Brid.) B.S.G.
Figure 27, D–E

6. TUERCKHEIMIA Bruch

Tuerckheimia angustifolia
(Saito) Zander
Figure 27, F

Gymnostomum angustifolium Saito
Molendoa sendtneriana (B.S.G.) Limpr. (in the sense of Breen, 1963).

Habitat: Damp calcareous rock.
Range: Asia; Alaska; southeastern United States; northeastern Mexico.
Gulf Coast Distribution: Florida.

This species is known along the Gulf Coast only from Jackson County, Florida, where it has been collected several times from limestone in the Florida Caverns State Park area. Its narrow leaves, calcareous habitat, plane leaf margins, and fragile leaf tips make it fairly distinctive. It was formerly known in the United States under the name *Molendoa sendtneriana* (B.S.G.) Limpr., another species.

7. TRICHOSTOMUM Bruch

Trichostomum molariforme
Zander
Figure 27, G

Plants small, yellowish to brownish green, ± glossy when dry due to the conspicuous shiny dorsal surface of the costa, in low wirey sods, the individual stems often bushlike because the longest leaves are aggregated toward the stem tips; stems erect, mostly simple; leaves loosely contorted and curved at tips when dry, stiffly erect-spreading when wet, linear-lanceolate from a broader hyaline base, mostly 1.5–2.5 mm long; upper cells very obscure, thick-walled, densely bulging-papillose, lower cells pale, smooth, rectangular, sometimes porose at base, not forming a V-shaped region; margins plane or slightly recurved above, somewhat irregular above; costa shortly excurrent into a stout pale point. Apparently dioicous; sterile in our area but some plants bear archegonia.

Habitat: A calciphile; on limestone and old mortar.
Range: Southern U.S.; Caribbean; Mexico; Guatemala.
Gulf Coast Distribution: Louisiana.

This attractive little moss, only recently recognized as a new species, is known in our area from only a single specimen collected at Fontainebleau State Park in St. Tammany Parish, Louisiana. It grew on old mortar between bricks on a ruined building. The bush-like appearance of the individual plants and the glossy dorsal surface of the costa make this species easy to recognize.

TRICHOSTOMUM TENUIROSTRE (Hook. & Tayl.) Lindb. [*T. cylindricum* (Bruch) C.M.] has been reported from our area in Texas. However, this is a moss of the interior of North America, in cool uplands, and I doubt that it actually occurs in our range. I have not seen any specimens from the Gulf Coast region.

8. TORTELLA (Lindb.) Limpr.

Plants mostly medium-size, dull, green to glaucous-green, yellowish or brownish, mostly in low turves or tufts; stems erect, simple or forked; leaves mostly strongly contorted when dry, erect-spreading when moist, oblong to linear-lanceolate or subulate, acute to obtuse, often apiculate, plane to concave; margins plane to erect or incurved, sometimes undulate; upper cells mostly very obscure, small, densely pluripapillose, subquadrate to hexagonal, basal cells conspicuously enlarged, hyaline, forming a distinct V-shaped border with the green upper cells; costa strong, percurrent to excurrent, smooth and shiny dorsally, conspicuous in the dry plants. Setae elongate; capsules mostly cylindric and erect; peristome mostly of 16 teeth split to the base into slender divisions, long, twisted; operculum long-conic to conic-rostrate; calyptra cucullate.

Two species occur along the Gulf Coast; one of them, *T. humilis*, is common essentially throughout our area. The genus is easily recognized by the conspicuous region of hyaline basal cells forming a V-shaped line where it meets the small, obscure, green cells of the upper part of the leaf. Compare also *Weissia jamaicense*.

Plants of varied habitats and substrates, common; leaves plane above, apex not cucullate; upper margins plane to erect, not incurved; median cells very obscure, 4.5–8.5 μm diameter; monoicous; peristome teeth twisted 2–3 times	1. T. HUMILIS
Plants coastal, on sandy and calcareous substrates, rare; leaves concave above, apex often cucullate; median or upper margins, or both, ± incurved; median cells distinct, 9.5–13 μm diameter; dioicous; peristome teeth straight	2. T. FLAVOVIRENS

Plants small, dull, ± glaucous-green, gregarious or more commonly forming low, dense turves; stems mostly short; leaves strongly contorted when dry, oblong to linear or broadly linear, sometimes acuminate, plane or somewhat concave above; apex acute, apiculate; median and upper cells very obscure, 4.5–8.5 μm diameter, very papillose, lower cells enlarged, rectangular, hyaline, extending further up along the margins than at the costa; margins finely papillose-roughened, plane to erect, sometimes undulate; costa yellowish, mostly short-excurrent into the yellowish apiculus. Frequently fertile; monoicous; setae yellowish above.

1. Tortella humilis (Hedw.) Jenn.
Figure 28, A–B

Habitat: Tree bases, bark, stumps, logs, rock and soil in mesic to dry forests.
Range: Europe; northern Africa; British Columbia; eastern North America; Mexico; Haiti; Jamaica; Brazil.
Gulf Coast Distribution: Throughout.

Tortella humilis is rather common along the Gulf Coast and is easy to recognize in both the field and lab. Its contorted leaves (when dry) with plane to erect margins and V-shaped areas of hyaline basal cells are distinctive. The hyaline base may occupy up to one-third of the leaf length. See comments under *T. flavovirens*.

Similar in habit and other respects to *T. humilis* but easily distinguished as stated in the key. Probably not fruiting in our area.

2. Tortella flavovirens (Bruch ex F. Müll.) Broth.
Figure 28, C–D

Habitat: Shelly sand, coquina, old concrete and brick walls, limey soil; in coastal areas near salt water.

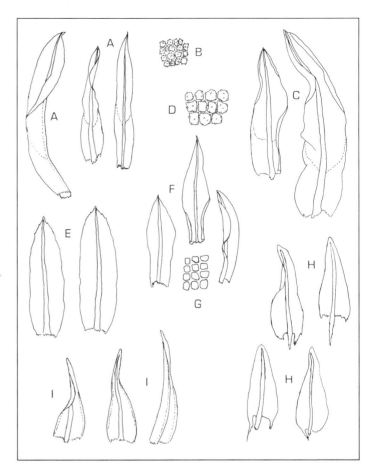

Fig. 28. *A–B:* **Tortella humilis.** *A,* leaves (× 15). *B,* median cells (× 231). *C–D:* **Tortella flavovirens.** *C,* leaves (× 15). *D,* median cells (× 231). *E,* **Hyophila involuta,** leaves (× 15). *F–G:* **Barbula agraria.** *F,* leaves (× 15). *G,* median cells (× 231). *H,* **Barbula tophacea,** leaves (× 15). *I,* **Barbula fallax,** leaves (× 15).

Range: Europe; North Africa; southeastern United States.
Gulf Coast Distribution: Alabama; Texas.

This moss is very rare in our area. The larger, distinct cells, often cucullate leaf tip, and incurved margins, as well as the maritime habitat, separate it easily from the very common *T. humilis.* I have seen specimens from our area from an old brick wall at Fort Gaines, Dauphin Island, Alabama, and from a sandy-shelly site in scrub on Goose Island, Aransas County, Texas.

9. HYOPHILA Brid.

Hyophila involuta (Hook.) Jaeg. & Sauerb.
Figure 28, E

Plants small, dull, dark green to brownish or black, in low, dense to loose turves, often bearing stalked, multicellular, echinate, polymorphous gemmae in upper leaf axils; stems erect, mostly simple; leaves often crowded at stem tips, oblong to spatulate, inrolled and somewhat contorted when dry, apiculate; cells small, quadrate, bulging ventrally, smooth to obscurely papillose dorsally, mostly in conspicuous transverse rows in upper part of leaf; margins plane, entire to serrulate or coarsely and irregularly toothed above; costa ending in apiculation. Not producing sporophytes in our area.

Habitat: Calcareous rocks and sandstone in moist, shaded sites.

Range: Europe; Australia; Asia; North America to South America.

Gulf Coast Distribution: Alabama.

This moss is usually easily recognized by its inrolled leaves, often coarsely toothed upper margins, and gemmae. Some specimens lack gemmae and some have entire margins. It is most common on calcareous substrates.

10. BARBULA Hedw.

Plants small to medium-size, sometimes tiny, dull, mostly in low turves or tufts, green to dark green, brownish, sometimes reddish, occasionally bearing axillary gemmae; leaves erect and straight or more often contorted when dry, variously lanceolate, sometimes acuminate, acute to obtuse, often apiculate; margins mostly revolute, at least below; upper cells rounded-quadrate, mostly obscure and papillose, rarely smooth, lower cells usually smooth, rectangular, hyaline. Setae elongate, mostly red; capsules oblong to ovoid or cylindric; peristome mostly of 16 slender teeth split to the base and twisted together, sometimes irregular and straight; spores smooth; operculum mostly long and bluntly rostrate; calyptra cucullate.

Five species of *Barbula* occur along the Gulf Coast but only one, *B. indica*, is at all common. Two species, *B. agraria* and *B. tophacea*, are ± strict calciphiles and are restricted to calcareous rocks (or old concrete). Plants of *Barbula*, in general, prefer calcareous substrates and can be recognized by the usually recurved leaf margins, twisted peristomes, and obscure, papillose cells. However, *B. tophacea* has a straight peristome and *B. agraria* has plane leaf margins and nonpapillose cells.

1. Leaf margins plane; cells smooth dorsally, bulging ventrally, not papillose	1. B. AGRARIA
1. Leaf margins recurved to revolute, at least below; cells conspicuously papillose	2
2. Cells of ventral surface of costa linear-oblong	3
2. Cells of ventral surface of costa quadrate to short-oblong	4
3. Leaves decurrent, obtuse, not acuminate; peristome teeth straight, irregular	2. B. TOPHACEA
3. Leaves not decurrent, both acute and acuminate; peristome teeth twisted	3. B. FALLAX
4. Axillary gemmae usually present, often abundant; leaves with short, small apiculation; plants weedy, frequent in disturbed sites	4. B. INDICA
4. Axillary gemmae lacking; leaves with stout apiculation; plants usually in natural habitats	5. B. UNGUICULATA

Plants tiny, virtually stemless, green to dark green, sometimes reddish; leaves straight and stiffly erect to somewhat contorted when

1. Barbula agraria Hedw.
Figure 28, F–G

dry, erect-spreading when moist, oblong to lanceolate or spatulate, obtuse to acute, apiculate; upper cells quadrate, mammillose ventrally, smooth dorsally, lower cells rectangular, smooth, hyaline; margins plane, entire to irregularly serrulate; costa ending in leaf apex or shortly excurrent into the apiculation. Setae red; capsules cylindric; peristome long and twisted; operculum bluntly long-rostrate.

> **Habitat:** Limestone; walls and old masonry.
> **Range:** Southern United States; tropical America.
> **Gulf Coast Distribution:** Florida to Texas.

Barbula agraria is common on limestone in humid regions and scarce elsewhere; it is much more common in the eastern part of its Gulf Coast range than it is to the west. It is usually found with sporophytes. The habitat; long, twisted peristome; smooth leaves with plane margins; and small size are distinctive.

2. Barbula tophacea (Brid.) Mitt.
Figure 28, G

Didymodon tophaceus (Brid.) Lisa

Plants small to medium-size, dull, green, usually encrusted with limey material in lower parts, in dense turves or cushions; stems erect, mostly simple; leaves ± contorted when dry, spreading when moist, usually strongly decurrent, ovate-lanceolate, mostly obtuse; margins recurved below, entire; upper cells distinct, quadrate to rounded-quadrate, papillose (papillae sometimes low), lower cells grading into upper ones, short-rectangular; costa ending below leaf tip. Probably not producing sporophytes in our area; capsules cylindric; peristome teeth straight, irregular.

> **Habitat:** Seeping, calcareous rock and soil.
> **Range:** Europe; Siberia; Japan; northern Africa; North and
>   South America.
> **Gulf Coast Distribution:** Texas.

Barbula tophacea is fairly easy to recognize because of its habitat, recurved leaf margins, decurrent leaves, costa ending below the leaf tip, and elongated cells on the ventral surface of the costa. It is a tufa-former, precipitating limey substances around the plants and thus becoming "permanently fossilized" (Grout 1928–1945, vol. 1, p. 190). *Gymnostomum aeruginosum*, also a calciphile, is somewhat similar but smaller, and the leaves are narrower, not decurrent, and usually lack recurved margins; also, plants of *G. aeruginosum* are often ± glossy, in contrast to the dull appearance of *Barbula tophacea*. This species is often treated in the genus *Didymodon*, because of its straight, irregular peristome.

3. Barbula fallax Hedw.
Figure 28, I

Plants mostly small, dull, dark green to brownish, in tufts or sods; leaves slightly contorted when dry, erect-spreading when moist, ovate-lanceolate, acuminate, acute, not decurrent; margins revolute below or nearly to the leaf tip; upper cells papillose, thick-walled, rounded-quadrate, lower cells rectangular, smooth, grading into the upper cells; costa ending in the leaf tip. Setae elongate; capsules cylindric; peristome teeth twisted.

Habitat: Calcareous soil in ± moist sites, especially road-cut banks.

Range: Widespread in the Northern Hemisphere.

Gulf Coast Distribution: Louisiana and Texas.

This little moss is rare along the Gulf Coast. It can be distinguished from *B. tophacea* by its acute and acuminate, nondecurrent leaves.

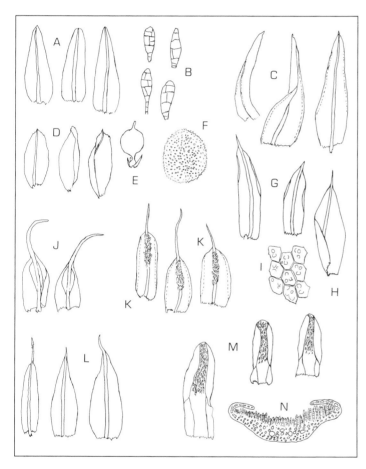

Fig. 29. *A–B:* **Barbula indica.** *A,* leaves (× 15). *B,* gemmae (× 91). *C,* **Barbula unguiculata,** leaves (× 15). *D–F:* **Acaulon muticum.** *D,* leaves (× 15). *E,* capsule (× 15). *F,* spore (× 231). *G–I:* **Phascum cuspidatum.** *G,* stem leaves (× 15). *H,* perichaetial leaf (× 15). *I,* median cells from stem leaf (× 231). *J,* **Pterygoneurum ovatum,** leaves (× 15). *K,* **Crossidium crassinerve,** leaves (× 15). *L,* **Desmatodon plinthobius,** leaves (× 15). *M–N:* **Aloina hamulus.** *M,* leaves (× 15). *N,* cross section of leaf (× 55).

Plants dull, yellowish to brownish green, sometimes appearing glaucous, in thin to dense low sods or sometimes gregarious, mostly bearing abundant, small, axillary, reddish multicellular gemmae; stems red, mostly simple; leaves contorted when dry, 1.5–2 mm long, oblong to oblong-lanceolate, sometimes acuminate, apex blunt, usually apiculate by a sharp, hyaline cell; upper cells very obscure, densely papillose, ± quadrate, lower cells rectangular, smooth, hyaline; margins revolute in lower ½–⅔ of leaf. Costa

4. **Barbula indica** (Hook.) Spreng. in Steud.
Figure 29, A–B

Barbula cruegeri Sond. ex C.M.
Barbula cancellata C.M.

strong, ending in leaf apex or shortly excurrent. Setae red; capsules yellowish, cylindric; peristome teeth long and slender, twisted, fragile and falling quickly; spores smooth; operculum long, bluntly rostrate.

Habitat: Calcareous rock and soil in ± disturbed areas; roadsides.
Range: Widespread, mostly in warmer parts of the world.
Gulf Coast Distribution: Florida to Texas.

This is a common and weedy moss over most of the Gulf Coast, particularly in sites which are in a more or less permanently disturbed state, such as roadsides, parks, paths, and campuses. In southern Louisiana it is a frequent associate, in disturbed sites, of *Weissia controversa*, *Tortula rhizophylla*, *Physcomitrium pyriforme*, and *Bryum argenteum*. Sporophytes are infrequently produced but the plants almost always bear the characteristic gemmae, often in great abundance. When gemmiferous plants are dissected in water on a microscope slide, the gemmae fall from the leaf axils and scatter on the slide like grains of sand.

Barbula indica was formerly known in the southern United States and the American tropics under the name *B. cruegeri*, and more recently as *B. cancellata*. See comments under *B. unguiculata*.

5. Barbula unguiculata Hedw.
Figure 29, C

Very similar in most respects to *B. indica*; differing mainly in the longer and stouter apiculation, the costa more regularly short-excurrent, the leaves more acuminate and acute, and the lack of gemmae.

Habitat: Banks of streams and rivers; old fields; roadsides.
Range: Cosmopolitan.
Gulf Coast Distribution: Florida to Texas; not yet reported from coastal areas of Mississippi and Alabama but probably present.

Saito (1975) and Zander (1979) point out additional differences between *B. indica* and *B. unguiculata*, for example, in the peristome teeth (10 μm thick in *B. indica*, ca 15 μm in *B. unguiculata*) and in the papillosity of the dorsal surface cells of the costa (doubly prorulose in *B. indica*, seriate-papillose in *B. unguiculata*). Such qualities are difficult to evaluate, and the most practical way to distinguish these two taxa in our area is by the presence or absence of the gemmae, correlated, of course, with the other characters. The two species also differ somewhat in habitat; *B. unguiculata* tends to occur, in our area, in relatively little-disturbed sites, such as stream and river banks, in contrast to the more weedy habitats of *B. indica*. *Barbula unguiculata* is rather uncommon along the Gulf Coast, whereas *B. indica* is very frequent.

Acaulon is a small genus of minute mosses occupying rather dry, open, often disturbed sites. The plants are rarely abundant and seldom collected. They are usually found only by careful search of likely sites because they are so small that they escape detection by all but the most ardent collector. Two species have been reported from our area, but it is likely that only one is actually present.

Plants tiny, bulbiform, not angled, virtually stemless, ca 1.5 mm tall, yellowish to brownish; leaves concave, broadly ovate, cuspidate, irregularly dentate to entire at apex; cells short-rhombic, smooth; upper margins erect to closely recurved; costa ending in leaf apex to short-excurrent and forming the reflexed apiculation. Cleistocarpous; setae shorter than the capsules; capsules immersed, globose, ± bluntly apiculate; spores smooth or warty; calyptra minute, consisting mostly of the neck of the archegonium.

> **Habitat:** Bare soil in ± open or brushy sites; old fields, roadsides.
> **Range:** Eastern North America.
> **Gulf Coast Distribution:** Mississippi to Texas.

This minute moss is seldom collected but is probably frequent all along the Gulf Coast. It fruits during the months from winter to early spring. The bulblike plants with smooth, concave leaves, are easy to recognize. Plants with smooth spores represent the variety *rufescens* (Jaeg.) Crum; those with papillose-warty spores represent the variety *muticum*. I have seen the smooth-spored form only from southern Texas, while the rough-spored form occurs in eastern Texas and Louisiana and probably to the east as well.

Acaulon schimperianum (Sull.) Sull. ex Sull. & Lesq. has been reported from our area, from Harris County, Texas. However, I have not seen any specimens of this moss from our area and doubt that it really occurs in our range. It differs from *A. muticum*, among other ways, in that the leaves have a long, slender acumination, and the cells in the upper part of the leaf are papillose dorsally.

Plants very small, yellowish to brownish green, loosely to densely gregarious; stems erect, very short; leaves ± contorted-twisted when dry, mostly 2–3 mm long, oblong-lanceolate, acute; upper cells mostly quadrate, smooth, or more often papillose with C-shaped and circular papillae, lower cells hyaline, lax, smooth, short-rectangular; margins mostly recurved at midleaf only; costa excurrent into a short or long, stout, brownish awn. Cleistocarpous; setae very short; capsules immersed, ovoid to globose, apiculate; spores finely papillose to tuberculate; calyptra small, cucullate.

> **Habitat:** Calcareous soil in ± open or brushy areas.
> **Range:** Europe; northern Africa; Caucasus; North America.
> **Gulf Coast Distribution:** Texas.

11. ACAULON C.M.

Acaulon muticum
(Hedw.) C.M.
Figure 29, D–F

12. PHASCUM Hedw.

Phascum cuspidatum Hedw.
Figure 29, G–I

This species is not common in our area. The leaf cells vary from essentially smooth to highly papillose. The immersed, cleistocarpous capsules and awned leaves with partially revolute margins are distinctive.

13. PTERYGONEURUM Jur.

Pterygoneurum ovatum
(Hedw.) Dix.
Figure 29, J

Plants tiny, yellowish green to brownish, sometimes hoary, often virtually stemless; leaves concave, ovate to spatulate, 1–1.5 mm long, in upper part with 2–4 prominent lamellae borne ventrally on the costa; cells isodiametric, smooth; margins erect; costa excurrent into a white hair-point shorter or longer than the lamina. Setae short to elongate; capsules emergent to exserted, ovoid, erect, furrowed when dry and empty; peristome lacking; spores papillose; operculum rostrate; calyptra cucullate.

> **Habitat:** Soil in dry, ± open sites.
> **Range:** Europe; northern Africa; western Asia; North America.
> **Gulf Coast Distribution:** Southern Texas.

This distinctive little moss, reported from Starr County, Texas, must be very rare in our area. The lamellose costa and hair-point are distinctive. *Pterygoneurum subsessile* (Brid.) Jur., differing, among other ways, in having a mitrate calyptra and immersed to emergent capsules, might also occur in our range in Texas. I have not seen the specimen upon which the Texas report was based, and it is possible that this species does not occur in our flora. It is not attributed to Texas by Crum and Anderson (1981).

14. CROSSIDIUM Jur.

Crossidium crassinerve
(De Not.) Jur.
Figure 29, K

Plants tiny, gregarious, reddish brown; stems very short; leaves wide-spreading when moist, lingulate, *ca* 1 mm long; cells smooth; margins revolute; costa bearing short, ventral filaments in upper part of leaf, excurrent into a long, white hair-point. Probably not producing sporophytes in our area; setae elongate; capsules ovoid-cylindric; peristome yellowish, twisted; spores smooth to papillose; operculum conic to rostrate.

> **Habitat:** Soil and rock in dry, ± open or brushy sites.
> **Range:** Europe; northern Africa; India; United States; Mexico.
> **Gulf Coast Distribution:** Southwestern Texas.

This little moss is known in our area only from southwestern Texas, but it occurs also in western Texas and elsewhere in the United States beyond the range of this manual. According to Delgadillo (1975) our plants belong to the variety *crassinerve*. The small plants with their short leaves bearing ventral filaments and a hair-point, and with revolute margins, are easy to identify.

15. DESMATODON Brid.

Desmatodon plinthobius Sull. & Lesq.
Figure 29, L

Plants small, closely gregarious, yellowish green to brownish, sometimes ± glaucous; stems erect, often forked; leaves mostly oblong, 1–2 mm long, obtuse; upper cells obscure, densely papillose, lower ones rectangular and hyaline, smooth; margins mostly revolute from hyaline base to leaf apex; costa excurrent into a hyaline hair-point sometimes as long as the lamina. Commonly fertile; setae elongate; capsules cylindric, erect and nearly symmetric; peristome teeth pale, irregular, very papillose; spores smooth; operculum bluntly rostrate; calyptra cucullate.

Habitat: Mostly calcareous substrates; soil, rock, old concrete and mortar, walls.
Range: Southeastern United States.
Gulf Coast Distribution: Florida to Texas.

This species is fairly common in our area, especially on old, man-made substrates such as tombstones and old brickwork. Human activities have surely increased the range and population of this plant by providing increased habitat for it. It is not rare, however, on native calcareous substrates. Whitehouse and McAllister (1954) list sandstone, as well as limestone, as a substrate. When sporophytes are present, as is usually the case, this moss is easy to recognize because of the short, irregular peristome and the long hair-points on the leaves.

Plants tiny, loosely or densely gregarious, dark green to reddish brown; stems erect, mostly simple; leaves lingulate, wide-spreading to almost squarrose when moist, *ca* 1.5 mm long, apex cucullate, margins broadly infolded; costa broad, densely covered ventrally with small, erect, branched filaments. Probably not producing sporophytes in our area; setae elongate; capsules erect, cylindric; peristome yellowish, straight or twisted; spores papillose; operculum conic to rostrate.

Habitat: Calcareous soil in ± dry, open sites.
Range: Southern United States; Mexico; Guatemala; El Salvador.
Gulf Coast Distribution: Louisiana.

This very interesting little moss is known in our area only from a single station in West Feliciana Parish, Louisiana, where it grows on loess on a roadside bank. The tiny plants with the costa densely covered with filaments are unmistakable. The stems are rather fleshy and break easily, the resulting fragments as well as detached leaves probably serving in vegetative reproduction.

Plants small to medium-size, green to dark green or brownish, scattered or tufted; stems ± erect; leaves mostly twisted around the stem when dry, lingulate to spatulate, wide-spreading to recurved when moist, mostly obtuse; upper cells quadrate to hexagonal, mostly papillose and obscure, sometimes smooth and pellucid; margins mostly revolute, at least in part, sometimes plane; costa mostly percurrent to excurrent, often forming an awn. Not producing sporophytes in our area.

16. ALOINA Kindb.

Aloina hamulus (C.M.) Broth.
Figure 29, M–N

17. TORTULA Hedw.

Our species of *Tortula* are small, inconspicuous plants not likely to be noticed in the field unless specifically sought. Two species occur along the Gulf Coast.

Plants growing on the bark of trees (rarely on rocks); cells distinctly pluripapillose; leaflike propagula commonly present in upper leaf axils .... 1. T. PAGORUM

Plants growing on soil or tree bases; cells smooth; propagula lacking in leaf axils .... 2. T. RHIZOPHYLLA

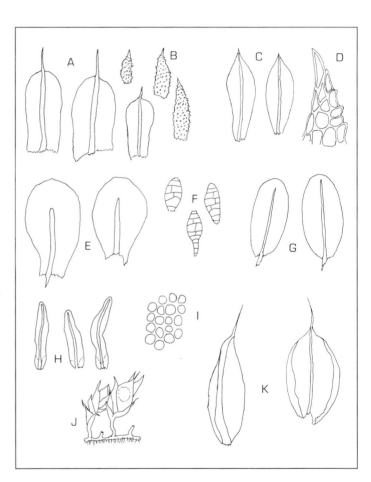

Fig. 30. *A–B:* **Tortula pagorum.** *A,* leaves (× 15). *B,* brood bodies (× 55). *C–D:* **Tortula rhizophylla.** *C,* leaves (× 15). *D,* leaf tip (× 231). *E–F:* **Gymnostomiella orcuttii.** *E,* leaves (× 55). *F,* gemmae (× 55). *G,* **Splachnobryum obtusum,** leaves (× 15). *H–I:* **Luisierella barbula.** *H,* leaves (× 15). *I,* median cells (× 231). *J–K:* **Lorentziella imbricata.** *J,* habit (× 3). *K,* leaves (× 15).

1. **Tortula pagorum** (Milde) De Not.
Figure 30, A–B

Plants tiny to small, dark green to brownish, scattered or tufted, usually bearing abundant, small, multicellular, papillose, elliptical, leaflike propagula in upper leaf axils; stems short; leaves slightly twisted when dry, mostly 1.5–2 mm long, broadly oblong-lingulate, rounded to slightly emarginate at apex; margins plane, ± crenulate-papillose; upper cells densely papillose with C-shaped and circular papillae, lower cells inside the margins ± hyaline and rectangular; costa excurrent into a short or long, stout, brownish to hyaline awn.

Habitat: Bark of hardwood trees in ± open, dry sites; rarely on rock.

Range: Europe; Asia; Australia; eastern United States; northeastern Mexico.

Gulf Coast Distribution: Mississippi to Texas.

This inconspicuous little moss is probably rather common over much of the Gulf Coast but is not often collected because it is so small. In parts of Texas it is common on mesquite bark. Like *T. rhizophylla*, this species seems to be associated in the United States with human activity and could be an introduction. Sporophytes are known only from Australia; our plants are either sterile or archegoniate. The habitat and propagula are characteristic.

Plants small, yellowish to brownish green, gregarious to scattered, bearing subterranean, brownish, multicellular gemmae (tubers) on the rhizoids; leaves ± incurved when dry, not twisted around the stem, lanceolate to spatulate, *ca* 1.3 mm long, acute, terminating in a sharp, brownish, often reflexed cell; margins plane, crenulate; upper cells pellucid, bulging-smooth to obscurely unipapillose, walls irregularly thickened, lower cells larger, lax, rectangular; costa ending below apex to percurrent.

2. Tortula rhizophylla (Sak.) Iwats. & Saito
Figure 30, C–D

Habitat: Bare soil in areas subject to disturbance; parks, campuses, fields; also on rain-splashed soil on walls and tree bases.

Range: England; Japan; Hawaii; Mexico; Bolivia; Italy; Australia; southern United States.

Gulf Coast Distribution: Louisiana.

This little moss can be recognized, even with a hand lens, by the sharp, brown, reflexed terminal cell of the leaf. It is known only from sterile and archegoniate plants throughout its range; the rhizoidal tubers serve to propagate the plants. Its occurrence seems to be correlated with human activity, as is also the case for *T. pagorum*. *Tortula rhizophylla* very likely occurs all along the Gulf Coast and should be sought in bare areas in parks, lawns, campuses, cemeteries. It is fairly common in southern Louisiana. The plants are usually partially buried in the fine soil in which they grow.

TORTULA MURALIS Hedw., a species with strongly revolute leaf margins, conspicuous awns, and densely papillose leaf cells, was reported long ago on the Gulf Coast, from New Orleans. However, I have not seen any Gulf Coast specimens of it and doubt that it really occurs in our area. It is a species inhabiting calcareous rocks.

18. GYMNOSTOMIELLA Fleisch.

Gymnostomiella orcuttii Bartr.
Figure 30, E—F

Plants small and slender, densely gregarious and forming low, soft turves, green to brown, sometimes with abundant, brown, multicellular gemmae in the axils of the lower leaves; stems short, erect, brown; leaves small, broadly short-lingulate to spatulate, very obtuse, less than 0.5 mm long; margins plane, papillose-crenate; upper cells ± quadrate, bearing one to several large, blunt papillae on both surfaces, lower cells larger, lax and rectangular; costa ending well below leaf apex, often unequally forked near the tip. Not producing sporophytes in our area.

> **Habitat:** Moist limestone.
> **Range:** Haiti; Jamaica; Puerto Rico; southern United States; eastern Mexico.
> **Gulf Coast Distribution:** Florida.

This delicate little moss is very pretty to see under the microscope. It resembles very much a small *Splachnobryum obtusum*, except for the papillose cells. Although it is known in our area only from Jackson County, Florida, there is no reason why it should not turn up eventually on calcareous substrates elsewhere along the Gulf Coast.

19. SPLACHNOBRYUM C.M.

Splachnobryum obtusum (Brid.) C.M.
Figure 30, G

Plants small, soft, loosely to densely gregarious, often with red tinge; stems erect, mostly simple, reddish; leaves shriveled when dry, spatulate, very obtuse, mostly 1—1.5 mm long; margins plane and crenulate above, sometimes recurved below; cells smooth, rhombic to hexagonal above, larger, rectangular and lax below; costa ending from well below to in the leaf apex, sometimes unequally forked at tip and spurred below. Not producing sporophytes in our area.

> **Habitat:** Calcareous rocks and soil in moist, protected sites; old mortar and brickwork.
> **Range:** Tropical and subtropical America; southern and southwestern United States; Europe.
> **Gulf Coast Distribution:** Florida; Louisiana; Texas.

Splachnobryum obtusum, sometimes classified in the Splachnaceae, is fairly common on moist calcareous substrates along the Gulf Coast, but is usually inconspicuous because of its small size. When moist, it can be identified in the field with a hand lens by its habitat and obtuse leaves. Even where natural calcareous substrates are lacking, *S. obtusum* can often be found on tombstones and similar substrates. Dr. A. J. Grout's "448. *Splachnobryum Bernoulii*," of his "North American Musci Perfecti," was collected from "concrete mortar and brick around tombs, St. Vincent de Paul Cemetery, New Orleans, Louisiana" (see Sayre, 1971, for description of Grout's distributions of dried specimens of mosses such as this one). (*Splachnobryum bernoulii* C.M. is a name formerly in use for the plant known as *S. obtusum* now.) This species is typical of the numerous mosses widely distributed in low-lying areas throughout the American tropics and along the Gulf Coast of the United States.

Plants tiny, often matted with blue-green algae, green to brown or reddish; stems very short, simple; leaves ligulate to linear, obtuse, straight to flexuous-contorted when dry, erect-spreading when moist, 1–2 mm long; margins plane, crenulate; upper cells quadrate to rounded, prominently bulging, thick-walled, not papillose, lower cells sharply differentiated from upper, hyaline, enlarged, extending up margins and forming a V-shaped border; costa ending below leaf apex. Commonly producing sporophytes; setae elongate; capsules cylindric, symmetric or nearly so; peristome teeth short to long, papillose, irregular, straight or slightly twisted when dry; spores smooth; operculum slenderly rostrate; calyptra cucullate.

> **Habitat:** Limestone, in shaded sites.
> **Range:** Tropical America; southern United States; Bermuda.
> **Gulf Coast Distribution:** Florida.

This tiny moss is easily recognized by its bulging cells in the upper part of the leaf and the hyaline, enlarged basal cells forming a V-shaped border. The gametophytes are often ± hidden in a mat of blue-green algae leaving only the sporophytes visible. The plants are so small that they are often found only by discovery of the sporophytes protruding from a discolored limestone surface. *Luisierella barbula* is known from our area of the Gulf Coast only from Jackson and Washington counties in Florida; however, it is also known from several sites in central Texas and could well be found in limey areas along the Texas Gulf Coast.

20. LUISIERELLA Thér. & P. de la Varde

Luisierella barbula (Schwaegr.) Steere
Figure 30, H–I

GRIMMIACEAE Arnott

Plants mostly small, dark green to brown or blackish, in tufts or mats, often hoary due to hyaline hair-points on the leaves; stems ± erect, often branched; leaves concave to channeled or keeled, appressed when dry, oblong to lanceolate, acute, mostly with a usually prominent hyaline hair-point; upper cells small, ± quadrate, smooth; lower cells somewhat elongate, often ± thick-walled and sinuose-porose; margins (and lamina) often thickened, especially above, plane or recurved, or erect to incurved; costa single, well-developed. Setae terminal, short to elongate; capsules immersed to exserted, short and broad to oblong, smooth or ribbed; peristome single, the 16 teeth sometimes perforate or split above; operculum apiculate to rostrate; calyptra cucullate or mitrate, naked, not or obscurely plicate.

GRIMMIA Hedw.

Grimmia is a large genus of mostly rock-dwelling mosses, mainly with a northern distribution. Only two species are known to occur along the Gulf Coast, and neither is common in our area. *Grimmia* could be confused with *Orthotrichum*, which also has small, dark plants growing in tufts. However, *Orthotrichum* grows, at least in our area, exclusively on bark while *Grimmia* grows exclusively on rock. *Ptychomitrium*, which also may grow on rock, differs from *Grimmia* in lacking hair-points, and in the calyptra plicate, and the peristome teeth divided nearly to the base.

1. Upper leaves without hair-points; capsule immersed — 1. G. APOCARPA
1. Upper leaves with well-developed hair-points; capsule immersed or exserted — 2
   2. Leaves with one or both margins recurved, at least below; capsule immersed — 1. G. APOCARPA
   2. Leaves with margins plane to erect or incurved; capsule exserted — 2. G. LAEVIGATA

1. Grimmia apocarpa Hedw.
Figure 31, A–B

Plants in small tufts, dark green to brownish; leaves keeled or channeled above, mostly with a short hyaline or brownish awn which is often inconspicuous and sometimes lacking; one or both margins revolute, at least below, thickened above. Capsules immersed; columella falling with the operculum; calyptra cucullate.

> **Habitat:** Sandstone and limestone, including old concrete; often in rather dry, exposed sites.
> **Range:** Nearly cosmopolitan.
> **Gulf Coast Distribution:** Louisiana.

This species is known along the Gulf Coast only from a few sites in central Louisiana where sandstone and limestone exposures occur. The revolute margins and usually inconspicuous awn are distinctive.

2. Grimmia laevigata (Brid.)
Brid.
Figure 31, C

Plants slender, dark green to blackish or brownish, usually hoary, commonly in dense clumps or mats; leaves concave, mostly bistratose, not keeled or channeled, ovate-lanceolate, with long hyaline hair-points to as long as the lamina; margins plane to erect. Setae short; capsules shortly exserted or reached by the hair-points; calyptra mitrate.

> **Habitat:** Sandstone.
> **Range:** Nearly cosmopolitan.
> **Gulf Coast Distribution:** Florida; Texas.

This moss is known in our area from a single station in Washington County, Florida, and from several stations in Texas. Its long hair-points and concave leaves make it easy to recognize. It should turn up in central Louisiana, where there are sandstone exposures.

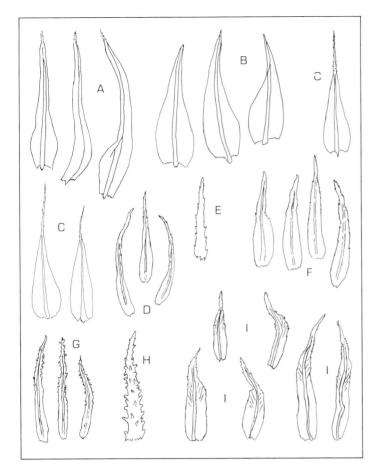

Fig. 31. *A–B:* **Grimmia apocarpa.** *A,* leaves with hyaline tips (× 15). *B,* leaves with concolorus tips (× 15). *C,* **Grimmia laevigata,** leaves (× 15). *D–E:* **Ephemerum crassinervium** var. **crassinervium.** *D,* leaves (× 21). *E,* leaf tip (× 55). *F,* **Ephemerum crassinervium** var. **texanum,** leaves, (× 21). *G–H:* **Ephemerum spinulosum.** *G,* leaves (× 21). *H,* leaf tip (× 55). *I,* **Ephemerum cohaerens,** leaves (× 15).

GRIMMIA OVALIS (Hedw.) Lindb. has been reported from Bastrop County, Texas, but probably does not occur in our area.

GIGASPERMACEAE Lindb.

Plants terrestrial, small to medium-size; secondary stems short, simple, erect from a pale, fleshy, subterranean rhizome, densely foliated above; leaves soft, small and remote below, crowded and large at stem tips; cells smooth, ± isodiametric to somewhat elongate; margins entire or weakly serrate; costa single or lacking. Sporophytes terminal, mostly immersed; peristome lacking; spores large; operculum sometimes not differentiated; calyptra small and fragile.

The Gigaspermaceae are a very small family of mosses, almost exclusively in the Southern Hemisphere. One genus, with a single species, occurs in the western part of our range.

LORENTZIELLA C.M.

Lorentziella imbricata (Mitt.)
Broth.
Figure 30, J–K

Plants medium-size, bulbiform, silvery green, gregarious; upper leaves closely imbricate, concave, broadly ovate, abruptly short-acuminate into a slender, hyaline, smooth or weakly toothed awn, the awn composed of thick-walled, elongate cells with very narrow lumens; upper part of leaves ± hyaline, lower part green; median and upper cells firm, quadrate to rhombic or short-rectangular, walls thickened at the common angles, cells in lower part of leaf ± quadrate; margins plane, entire or weakly serrate above; costa rather weak, scarcely extending into the awn. Usually fertile; antheridia in a leaf axil below the perichaetial leaves; sporophyte completely immersed; setae very short; capsules pale, ± spherical to bluntly ovate, sometimes bluntly apiculate; operculum not differentiated; spores brown, finely and irregularly granular, to ca 190 µm; calyptra tiny, campanulate, fugacious.

Habitat: Soil in open or brushy sites.
Range: North America (Texas); southern South America.
Gulf Coast Distribution: Texas.

This curious little moss is known from a fair number of stations in Texas, both inside and outside of our range. The bulbiform (when capsules are present) plants are fairly conspicuous, and the silvery green aspect, rhizomes, and immersed, cleistocarpous capsules are characteristic. The plants somewhat resemble those of *Funaria hygrometrica* or *F. flavicans*, before their setae begin to elongate. All other members of this family occur only in the Southern Hemisphere; it is interesting to speculate on how *L. imbricata* reached North America.

EPHEMERACEAE Schimp.

Plants tiny, ephemeral, scattered on the often abundant, ± persistent protonemata, virtually stemless; leaves few, the upper ones larger than the lower, linear to ovate-lanceolate, acuminate, margins entire to serrate-ciliate; cells oblong-rhombic, lax and clear, smooth, or papillose by projecting upper ends of cells; costa single or lacking, often weak. Mostly cleistocarpous but the upper half of the capsules sometimes separating along a line; capsules sessile, ovoid to globose, sometimes apiculate; spores large, papillose; calyptra small to minute, mitrate or consisting essentially of the remains of the archegonium.

Plants of this family have, as implied by the name, a short life span. The plants are fertile, and most conspicuous, in the cooler months and mostly disappear in the summer, presumably surviving as spores. The surest method for finding these mosses is to look for patches of the green, algalike protonemal mats on otherwise bare soil, and to check with a hand lens to see if the leafy, fertile plants are present. It is not unusual for two or more species of this family to grow together. These tiny mosses are not commonly collected, but this is only due to their inconspicuous nature, since at least several of our species are quite common.

Leaves mostly with an evident costa (some virtually ecostate), not particularly shriveled when dry; calyptra mitrate, covering ± half of the capsule; capsule not dehiscing	1. EPHEMERUM
Leaves essentially ecostate, shriveled when dry; calyptra very tiny, covering only the very tip of the capsule; capsule in several species dehiscing near the middle	2. MICROMITRIUM

1. EPHEMERUM Hampe

Leaves mostly with a distinct costa, entire to serrate or spinose, linear to lanceolate or ovate-lanceolate, acuminate, not particularly shriveled or contorted when dry. Capsules ovoid to globose, apiculate, indehiscent; calyptra mitrate, smooth to papillose, naked.

Three species of *Ephemerum* occur in our area; one of the species is represented by two varieties. *Ephemerum* is distinguished from *Micromitrium*, among other ways, by its larger calyptra and mostly costate leaves.

1. At least some leaves with ± prominent shoulders above; cells mostly smooth, in at least some leaves arranged in oblique rows from costa to margins at the shoulders	3. E. COHAERENS
1. Leaves mostly lacking shoulders; cells smooth or papillose, not arranged in oblique rows	2
2. Upper leaves mostly linear-setaceous, with spinose marginal cells mostly spreading to recurved at 45° or more	2. E. SPINULOSUM
2. Upper leaves narrow to broad, entire to toothed, the teeth spreading mostly less than 45°	1. E. CRASSINERVIUM

Leaves ovate- to oblong-lanceolate, to linear-setaceous, nearly entire to strongly toothed above, the teeth mostly not or only slightly spreading, without shoulders or with slight shoulders; cells mostly distinctly papillose. Capsules yellowish to brown.

1. Ephemerum crassinervium (Schwaegr.) Hampe
Figure 31, D–F

> **Habitat:** Exposed soil in low forests, ditch banks, roadsides, old fields.
> **Range:** Eastern North America.
> **Gulf Coast Distribution:** Florida to eastern Texas.

This species occurs along the Gulf Coast as two varieties: var. *crassinervium* (Fig. 31, D–E), with the leaves lanceolate to linear and lacking shoulders; and var. *texanum* (Grout) Bryan & Anderson (Fig. 31, F), with the leaves more broadly ovate- to oblong-lanceolate, often with evident shoulders, and ± abruptly subulate. Both varieties probably occur throughout our area, but neither is very common.

2. Ephemerum spinulosum
Bruch & Schimp. ex Schimp.
Figure 31, G–H

Leaves mostly linear-setaceous from a narrowly lanceolate or oblong base, usually conspicuously spinose on the upper margins with the teeth often spreading at 45° or more and sometimes recurved, mostly lacking shoulders; cells smooth or papillose above. Capsules very small, brown, usually scarce.

> **Habitat:** Exposed soil and mud in low forests, roadsides, drying sloughs and pond margins, and along streams.
> **Range:** Europe; eastern North America.
> **Gulf Coast Distribution:** Louisiana; eastern Texas.

This is the most common species of *Ephemerum* along the Gulf Coast. The spines on the leaves can often be seen with a hand lens.

3. Ephemerum cohaerens
(Hedw.) Hampe
Figure 31, I

Leaves ovate- to broadly ovate-lanceolate or more narrow, often abruptly acuminate-subulate, often with ± prominent serrate to spinose shoulders; cells ± smooth, in at least some leaves, those in the middle ⅓ of the leaf arranged in oblique rows from the costa to the margins. Capsules brown.

> **Habitat:** Exposed soil in forests and along streams and road banks, in moist sites.
> **Range:** Europe; eastern North America
> **Gulf Coast Distribution:** Louisiana; eastern Texas.

The often prominently shouldered leaves with the cells arranged in oblique rows are distinctive. This species is probably fairly common over much of our area.

EPHEMERUM SERRATUM (Hedw.) Hampe has been reported from our area, in Bastrop and Hidalgo counties, Texas. However, I have not been able to study the specimens upon which the reports are based and do not believe that the species occurs in our area. It is distinct from all others in the genus by its ecostate leaves; its larger calyptra will distinguish it from *Micromitrium*.

2. MICROMITRIUM Aust.
Nanomitrium Lindb.

Leaves ecostate, entire to serrate, lanceolate to ovate-lanceolate, ± contorted-shriveled when dry. Capsules globose, scarcely apiculate, some dehiscing near the middle; calyptra tiny, sitting on top of the capsule, consisting mostly of the neck of the archegonium.

Three species of *Micromitrium* have been reported from the Gulf Coast. The ecostate leaves and minute calyptrae are distinctive. Some *Ephemerum* species have the costa weak in the lower part of the leaf but the larger calyptrae will separate *Ephemerum* from *Micromitrium*.

1. Mature capsule dark brown to black; spores less than 40 μm diameter; stomata lacking on capsule wall   3. M. AUSTINII
1. Mature capsule yellow to orange-brown; spores more than 40 μm diameter; stomata present on capsule wall   2

2. Stomata only at base of capsule; leaves very narrowly lanceolate-acuminate, ± entire .... 1. M. WRIGHTII
2. Stomata only on upper half of capsule; leaves ovate-
   lanceolate, short-acuminate, serrulate above .... 2. M. MEGALOSPORUM

Leaves very slender, long acuminate, mostly entire. Capsules yellowish to orange-brown when mature; spores mostly about 60 μm
diameter, papillose.

1. Micromitrium wrightii
(C.M.) Crosby
Figure 32, A

Nanomitrium wrightii (C.M.)
Bryan & Anderson

> **Habitat:** Exposed soil and mud in drying sloughs and pond
> margins.
> **Range:** Southern United States; Cuba.
> **Gulf Coast Distribution:** Louisiana.

This little moss is known in the United States only from a few
collections from central and southern Louisiana, although it could
well be more widespread than these specimens indicate. Its very
slender leaves, mostly entire, distinguish it from *M. megalosporum.*
It is very difficult to see the stomata.

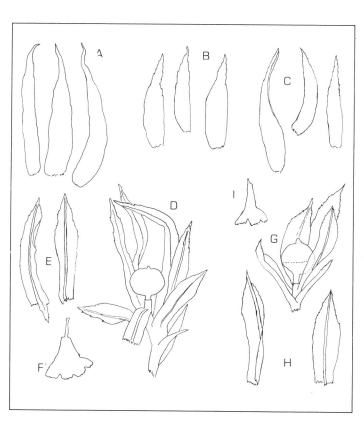

Fig. 32. *A,* **Micromitrium wrightii,**
leaves (× 16). *B,* **Micromitrium
megalosporum,** leaves (× 16). *C,*
Micromitrium austinii, leaves (×
16). *D–F:* **Physcomitrella patens.**
D, habit (× 15). *E,* leaves (× 15). *F,*
calyptra (× 16). *G–I:* **Aphanorrhegma serratum.** *G,* habit (× 15).
H, leaves (× 15). *I,* calyptra (× 30).

2. **Micromitrium megalosporum** Aust.
Figure 32, B

Nanomitrium megalosporum (Aust.) E. G. Britt.

Leaves rather broadly ovate-lanceolate, serrulate toward apex. Capsules orange-brown when mature, with stomata in upper part; spores mostly about 50–57 μm diameter, ± papillose.

> **Habitat:** Exposed soil and mud in low places; drying sloughs, ponds, lake margins.
> **Range:** Eastern United States; Cuba.
> **Gulf Coast Distribution:** Louisiana.

This species is known from several specimens from Louisiana but apparently has not been collected elsewhere in our area. Like most other members of the Ephemeraceae, it is probably more common than existing collections indicate. The broader leaves distinguish it from *M. wrightii*.

3. **Micromitrium austinii** Aust.
Figure 32, C

Nanomitrium austinii (Aust.) Lindb.

Leaves delicate, narrowly ligulate-lanceolate, serrulate toward apex. Capsules tiny, dark brown to black when mature, lacking stomata; spores mostly 27–35 μm diameter, papillose.

> **Habitat:** Exposed soil and mud in low places in wet forests.
> **Range:** Eastern United States.
> **Gulf Coast Distribution:** Louisiana.

This distinctive species is apparently rare along the Gulf Coast, where it has only been reported from Louisiana in our area. The small, very dark brown or black capsules and small spores distinguish it easily from *M. megalosporum* and *M. wrightii*.

FUNARIACEAE Schwaegr.

Plants tiny to medium-size, gregarious or tufted, light to yellowish green; stems erect, simple or forked, very short to ± elongate; leaves soft, increasing in size upwards, crowded above, mostly ± oblong-lanceolate, entire to serrate; cells smooth, mostly lax, large, rhombic to oblong above, ± rectangular below, often weakly differentiated on the margins; costa single, ending near the leaf tips to excurrent. Cleistocarpous or (mostly) stegocarpous; setae very short to elongate; capsules immersed to (mostly) long-exserted, erect and symmetric to inclined and strongly curved, often with a short neck; peristome lacking or single or double, of 16 teeth, segments of the endostome opposite the teeth, sometimes reduced and inconspicuous; operculum mostly differentiated, hemispheric to conic or apiculate; calyptra smooth, mitrate or cucullate, small to large and inflated, often lobed at base.

The Funariaceae constitute a rather small family of rather soft, terrestrial mosses. Most members are weeds of disturbed areas; some are commonly found on burned sites. The wide range of variation among the sporophytes of this family illustrates the difficulty of making classification systems for mosses; however, the gametophytes are rather uniform and characteristic throughout. Six genera, with 10 species, occur along the Gulf Coast, but only a few species are common.

1. Capsules long-exserted; peristome present	2
1. Capsules immersed to exserted; peristome lacking	3
2. Capsules curved and asymmetric; peristome double (segments sometimes low and inconspicuous)	5. Funaria
2. Capsules erect and symmetric; peristome single	6. Entosthodon
3. Capsules inoperculate, rupturing irregularly around the middle; exothecial cells lax and thin-walled; calyptra tiny, less than 0.5 mm wide at base	1. Physcomitrella
3. Capsules with well-defined operculum; exothecial cells mostly ± thick-walled or collenchymatous; calyptra small or large	4
4. Capsules immersed	5
4. Capsules emergent to exserted	6
5. Calyptra tiny, less than 0.5 mm wide at base; exothecial cells collenchymatous	2. Aphanorrhegma
5. Calyptra larger, much more than 0.5 mm wide; exothecial cells ± thick-walled but not collenchymatous	3. Physcomitrium sp.
6. Calyptra ± persistent, ± 4-angled; capsules emergent to short-exserted; seta about as long as capsule	4. Pyramidula
6. Calyptra deciduous, not 4-angled; capsules exserted; seta longer than the capsule	3. Physcomitrium sp.

1. PHYSCOMITRELLA B.S.G.

Physcomitrella patens (Hedw.) B.S.G.
Figure 32, D–F

Plants tiny, gregarious, sometimes with a reddish tinge; stems short; leaves ovate to narrowly lanceolate, toothed above. Setae very short; capsules immersed, globose, bluntly apiculate above or rounded, rupturing irregularly, exothecial cells lax and thin-walled; calyptra tiny, less than 0.5 mm wide, mitrate; spores papillose.

Habitat: River banks; disturbed soil in wet fields; pond margins.
Range: Europe; northern Asia; eastern North America.
Gulf Coast Distribution: Louisiana.

This little moss is very rare and is known in our area only from a single collection from near New Orleans. The cleistocarpous capsules with lax, thin-walled exothecial cells, and the tiny calyptra, are distinctive. See comments under *Aphanorrhegma serratum*.

2. APHANORRHEGMA Sull.

Aphanorrhegma serratum (W. J. Hook. & Wils. ex Drumm.) Sull.
Figure 32, G–I

Plants tiny, gregarious, virtually stemless; leaves oblong-lanceolate, serrate above. Setae very short; capsules immersed, globose, dehiscing (sometimes irregularly) around the middle, exothecial cells distinctly collenchymatous; peristome lacking; operculum hemispheric, apiculate or rounded; calyptra tiny, less than 0.5 mm wide, mitrate, lobed at base; spores papillose.

Habitat: On mud and silt banks along rivers; disturbed soil.
Range: Eastern United States.
Gulf Coast Distribution: Louisiana and Texas.

These beautiful little plants are only rarely collected because of their small size. They differ from *Physcomitrella* in the collenchymatous exothecial cells and from *Physcomitrium immersum* and *Pyramidula* in the minute calyptra. Also, in *Pyramidula* the capsules are emergent to short-exserted. Recently deposited mud along river banks is a prime place to seek this moss.

3. PHYSCOMITRIUM (Brid.) Fürnr.

Plants tiny to medium-size, ± glossy, green to yellowish green, gregarious to tufted; stems erect, mostly simple; leaves soft, shriveled when dry, mostly rather broadly oblong-lanceolate, entire to serrate; cells large and lax; costa ending in the leaf apex to shortly excurrent. Setae very short to elongate; capsules immersed to exserted, ± pyriform, with short neck; peristomes lacking; spores papillose; operculum convex, sometimes with a blunt or pointed, short or long apiculus; calyptra large, inflated, with long beak, split into lobes below at maturity.

Physcomitrium is a small genus of terrestrial mosses that occupy ephemeral habitats. The large calyptrae and erect, operculate capsules lacking peristomes make the genus easy to recognize. Three species occur along the Gulf Coast; only one of them, *P. pyriforme*, is common.

1. Plants minute; capsules immersed	3. P. IMMERSUM
1. Plants small to medium-size; capsules exserted	2
2. Plants small, delicate; seta 2–3 mm long; capsule broadly flared from the base when dry and empty, with age nearly saucer-shaped; rare	2. P. COLLENCHYMATUM
2. Plants medium-size; seta mostly much more than 3 mm long; capsules not flared or flared only at the mouth when dry, never flattened and saucer-shaped; common	1. P. PYRIFORME

1. Physcomitrium pyriforme
(Hedw.) Hampe
Figure 33, A–C

Plants small to medium-size, green to yellowish green, gregarious; leaves oblong-lanceolate, acuminate, nearly entire to serrate above; costa ending in apex to shortly excurrent. Setae mostly elongate, sometimes as short as 4–5 mm; capsules globose-pyriform, contracted when dry and empty or constricted below the flaring mouth; operculum bluntly apiculate.

Habitat: A common weed of disturbed soil; fields, gardens, lawns, forests, roadsides.
Range: Europe; northern Africa; Australia; North America; Mexico.
Gulf Coast Distribution: Florida to Texas.

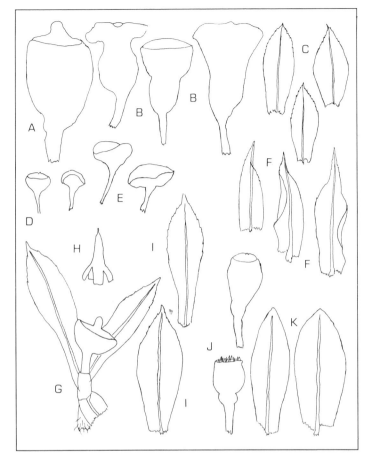

Fig. 33. *A–C*: **Physcomitrium pyriforme.** *A,* capsule with operculum (× 15). *B,* older, deoperculate capsules (× 15). *C,* leaves (× 15). *D–F*: **Physcomitrium collenchymatum.** *D,* capsule with operculum (× 15). *E,* older, deoperculate capsules (× 15). *F,* leaves (× 15). *G–I*: **Physcomitrium immersum.** *G,* habit (× 15). *H,* calyptra (× 15). *I,* leaves (× 15). *J–K*: **Entosthodon drummondii.** *J,* capsules, one with operculum (× 15). *K,* leaves (× 15).

This common and often abundant species is variable in several respects but is easy to recognize in general. It could be confused with *P. collenchymatum,* but differs in its larger plants, longer seta, and capsule flaring only at the mouth rather than from the base.

Plants rather small, green, gregarious; leaves obovate to oblong, acute, serrulate above; costa ending in leaf apex. Setae short, 2–3 mm long; capsules ± pyriform when young, expanding and flattening with age, sometimes becoming saucer-shaped; operculum bluntly apiculate.

Habitat: Mud and silt along streams; damp disturbed sites, including fields and around houses.
Range: Central and southern United States.
Gulf Coast Distribution: Louisiana.

This moss is surely more widely distributed along the Gulf Coast than is indicated by the handful of specimens. It is small and can be confused with *P. pyriforme*; see comments under the latter species.

2. **Physcomitrium collenchymatum** Gier
Figure 33, D–F

3. **Physcomitrium immersum**
Sull.
Figure 33, G–I

Plants tiny, yellowish green, loosely gregarious; stems often very short; leaves lanceolate, acute, bluntly serrate above. Setae very short; capsules globose to ± pyriform, immersed; operculum convex, bluntly apiculate; calyptra about 1 mm wide at base.

> **Habitat:** Recently exposed mud and silt along rivers and around ponds, reservoirs, and lakes.
> **Range:** Widespread in North America.
> **Gulf Coast Distribution:** Louisiana and eastern Texas.

This tiny moss is probably more common in our area than the few collections would indicate. It is similar to *Aphanorrhegma* but differs in having a much larger calyptra and in the operculum dehiscing from the upper part of the capsule rather than at midcapsule.

4. PYRAMIDULA Brid.

Pyramidula tetragona (Brid.) Brid.
Figure 34, A–D

Plants tiny, gregarious, essentially stemless; leaves ovate, slenderly acuminate, entire; costa often long-excurrent. Setae very short; capsules emergent to short-exserted, ± pyriform; peristome lacking; operculum convex, bluntly apiculate; calyptra strikingly 4-angled and covering the entire capsule, at least when young, split on one side; spores smooth.

> **Habitat:** Soil and rock in open areas.
> **Range:** Europe; northern Africa; central United States.
> **Gulf Coast Distribution:** Texas.

The strongly 4-angled calyptra and emergent to slightly exserted capsule are distinctive. *Pyramidula tetragona* is rare along the Gulf Coast, where it is known only from a few collections from southern Texas.

5. FUNARIA Hedw.

Plants small to medium-size, loosely to densely gregarious, green to yellowish green, stems mostly simple; leaves mostly contorted when dry, entire to serrate, the upper ones larger and sometimes forming bulbous clusters at the stem tips; cells large and lax; costa mostly ending in leaf apexes, sometimes excurrent. Setae elongate; capsules mostly curved and asymmetric, inclined to pendulous, often furrowed when dry, mouth oblique, annulus large and revoluble or lacking; peristome double, teeth 16, sometimes united at the tips and forming a delicate perforated disc, segments opposite the teeth, nearly as long as the teeth or much shorter, sometimes low and inconspicuous; spores smooth to papillose; operculum mostly convex; calyptra large, inflated, cucullate, with a long beak.

The long-exserted, asymmetric capsules, large, inflated calyptrae, and double peristomes make *Funaria* easy to recognize. Three species occur along the Gulf Coast.

1. Annulus present; spore-bearing portion of capsules furrowed or wrinkled when dry; exothecial cells nearly isodiametric or

oblong, width of lateral walls mostly less than width of lumens; leaves entire or slightly serrulate 2

1. Annulus lacking; spore-bearing portion of capsules essentially smooth when dry; exothecial cells short-linear, width of walls equal to or wider than width of lumens; leaves serrate above .... 3. F. SERRATA

  2. Endostome segments ¾ or more length of the teeth; spores 12–15 μm diameter .... 1. F. HYGROMETRICA

  2. Endostome segments less than ½ length of teeth or lacking; spores 20–30 μm diameter .... 2. F. FLAVICANS

Plants densely gregarious or tufted, sometimes scattered, gametophytes yellowish green, sporophytes yellowish to brown or dark red or purple; leaves mostly entire, usually crowded above and forming a bulblike tuft at the stem tip; costa subpercurrent to short-excurrent. Setae rather short to very long, 1–8 cm; capsules with annulus, very asymmetric, inclined to pendulous, slightly to strongly curved, furrowed when dry, mouth highly oblique; seg-

1. Funaria hygrometrica
Hedw.
Figure 34, E–F

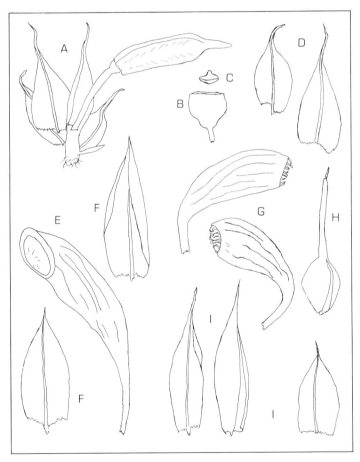

Fig. 34. *A–D*: **Pyramidula tetragona.** *A*, habit (× 15). *B*, capsule (× 15). *C*, operculum (× 15). *D*, leaves (× 15). *E–F*: **Funaria hygrometrica.** *E*, capsule (× 15). *F*, leaves (× 12). *G–I*: **Funaria flavicans.** *G*, capsules (× 15). *H*, calyptra (× 11). *I*, leaves (× 12).

ments of the endostome ⅔ or more as long as the teeth; spores mostly 14–20 μm.

Habitat: Disturbed sites on soil; burned areas; fields and roadsides.
Range: Nearly cosmopolitan.
Gulf Coast Distribution: Florida to Texas.

Funaria hygrometrica is a fairly common weed essentially throughout the Gulf Coast, where it produces sporophytes during the spring and summer, and sometimes even into the fall.

2. **Funaria flavicans** Michx.
Figure 34, G–I

Plants yellowish green, in dense tufts; leaves clustered above, often forming a bulblike aggregation at the stem tip, not contorted to strongly contorted when dry, entire; costa, at least in upper leaves, strongly excurrent into a sharp awn. Setae to 4 cm or more long; capsules with annulus, furrowed or wrinkled when dry, inclined to pendulous, slightly curved, mouth slightly oblique; segments of endostome about ¼ length of teeth, blunt, papillose, ± fused to the teeth below; spores 20–30 μm diameter.

Habitat: Disturbed soil; burned sites; roadsides.
Range: Eastern North America.
Gulf Coast Distribution: Florida to Texas.

Funaria flavicans is common along the Gulf Coast and is similar at first glance to the equally common *F. hygrometrica*. However, the mouth of the capsule in *F. flavicans* is less oblique than in *F. hygrometrica*. Under the microscope the large spores and short segments of the endostome distinguish it easily from *F. hygrometrica*.

3. **Funaria serrata** Brid.
Figure 35, A–B

Plants rather small, yellowish green, scattered to loosely gregarious; leaves clustered above, contorted when dry, broadly acute, serrate above, costa mostly ending a few cells below apex of leaf. Setae 1–2 cm long; capsules inclined, strongly curved when dry, spore-bearing portion smooth when dry, ca 1.5–2 mm long; segments nearly as long as teeth.

Habitat: Banks and disturbed soil; roadsides and low forests.
Range: Southeastern United States.
Gulf Coast Distribution: Florida to Texas.

These little plants are easy to recognize because of their size, serrate leaves, and strongly curved, smooth capsules. This species is probably common throughout our area, but the plants are not usually conspicuous because they tend to grow as scattered individuals or in loose aggregations, rather than in dense tufts as do some of the other species of *Funaria*. *Entosthodon drummondii* is a frequent associate of *Funaria serrata*.

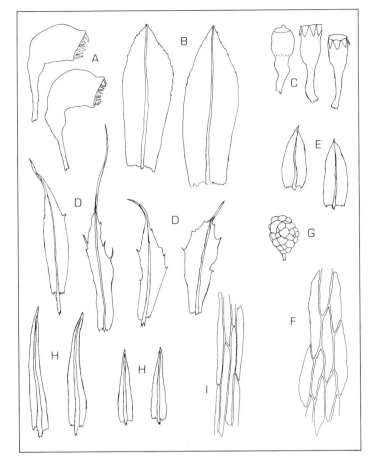

Fig. 35. *A–B:* **Funaria serrata.** *A,* capsules (× 15). *B,* leaves (× 15). *C–D:* **Splachnum pennsylvanicum.** *C,* capsules, one with operculum (× 15). *D,* leaves (× 15). *E–G:* **Pohlia apiculata.** *E,* leaves, (× 15). *F,* upper median cells (× 231). *G,* gemma (× 55). *H–I:* **Pohlia wahlenbergii.** *H,* leaves (× 12). *I,* upper median cells (× 231).

Funaria hygrometrica var. calvescens (Schwaegr.) Mont. has been reported from our area but apparently does not occur here. According to Crum and Anderson (1981) the Gulf Coast is not included in the range of the moss in North America, and I have not seen any convincing specimens from our area.

Funaria microstoma Bruch ex Schimp. was reported from our area (Newton County, Texas) by Whitehouse and McAllister (1954). However, the species does not occur in Texas according to the recent treatment by Fife (1979). Reports of F. americana Lindb. from eastern Texas are apparently based on misidentifications.

Plants sparsely (rarely densely) gregarious, yellowish green, rather small; stems short, simple; leaves mostly oblong-lanceolate, entire or slightly serrate above, apex rather blunt, sometimes acute; costa ending near leaf apex. Setae elongate; capsules erect and symmetric, pyriform, with an evident short neck; peristome single; operculum slightly convex, mostly with a low, blunt apiculation;

6. ENTOSTHODON Schwaegr.

Entosthodon drummondii Sull.
Figure 33, J–K

spores verrucose; calyptra cucullate, inflated, with a long beak, split on one side from the base.

Habitat: Soil; clearings and open forests in pine and mixed forest areas; dry savannas and roadsides.
Range: Southeastern United States.
Gulf Coast Distribution: Florida to eastern Texas.

Entosthodon is essentially a *Funaria* with the capsules erect and symmetric and a single peristome. It is fairly common, mostly in pine areas, but only rarely abundant. It often grows in grassy sites, where it is very inconspicuous; *Funaria serrata* is a frequent associate.

SPLACHNACEAE Grev. & Arnott

Plants small to medium-size, in soft cushions or sods; stems erect; leaves ovate-lanceolate to spatulate, margins entire or variously toothed, sometimes bordered with differentiated cells; cells smooth, lax, large, oblong to rectangular or rhombic; costa single, strong, ending below leaf apex to long-excurrent. Setae elongate; capsules mostly erect and symmetric, usually with a distinct neck, which is sometimes highly modified and colored; peristome single, the 16 teeth often associated in 2's or 4's, sometimes reflexed when dry; operculum convex to conic; calyptra small, mostly mitrate and naked.

Members of the Splachnaceae are mostly adapted to grow on organically enriched substrates, such as old dung and bones. The brightly colored capsule and neck of some members of the family appear to be a mechanism to attract insects for spore dispersal. Only one species occurs in our area.

SPLACHNUM Hedw.

Splachnum pennsylvanicum
(Brid.) Grout ex Crum
Figure 35, C–D

Tetraplodon pennsylvanicum
(Brid.) Sayre in Grout

Plants small, in low, soft cushions, dark to yellowish green, matted with red rhizoids below; stems short, erect, mostly simple; leaves long and slenderly lanceolate-subulate, shriveled when dry; cells large, clear, delicate, rectangular to oblong and rhombic, slightly larger in leaf base than above; margins entire to coarsely toothed-ciliate, marginal cells narrow, forming an indistinct border; costa strong, mostly long-excurrent and forming the slender, ± flexuous subula. Setae fleshy, pale, short; capsules with a usually prominent neck, orange to red; peristome teeth in 2's or 4's, reflexed when dry; operculum low, conic to hemispheric, bluntly apiculate; calyptra tiny, conic, 0.5 mm long.

Habitat: Dung of herbivores, old bones, organic soil, old concrete, swampy sites.
Range: Eastern North America.

Gulf Coast Distribution: Florida; Mississippi; Louisiana; Texas.

This species is very rare in our area but probably occurs all along the Gulf Coast from eastern Texas to Florida. It should be sought on old horse and cow dung in wet areas.

BRYACEAE Schwaegr.

Plants mostly in dense tufts or turves, yellowish green or darker; bulbils, axillary gemmae, or rhizoid tubers produced by some; stems erect, commonly forked, often with dense tomentum in lower parts; leaves ovate to lanceolate or narrower, rarely subulate, mostly imbricate, not much contorted when dry; cells smooth, mostly rhombic, sometimes longer and narrower; margins plane to recurved, entire to serrate, often bordered with narrow, elongate cells; costa single, strong, ending below leaf apex to strongly excurrent. Sporophytes mostly terminal; setae elongate; capsules mostly symmetric, inclined to pendulous, usually with conspicuous necks; peristome double; spores smooth to papillose; operculum short, convex to conic, umbonate to apiculate; calyptra cucullate.

The Bryaceae are a large family of mostly rather soft, ± glossy mosses, with a mainly terrestrial habitat. Four genera occur along the Gulf Coast, but only one of them, *Bryum*, is well represented and common in our area.

1. Upper leaves very long and narrow, linear-subulate, the acumination consisting mostly of the costa ..... 3. LEPTOBRYUM
1. Upper leaves broader, ovate to ovate-lanceolate or lanceolate, never subulate, with conspicuous lamina throughout 2
  2. Plants corticolous in our area, usually on the bark of live oak and magnolia; leaves spirally twisted around the stem when dry; axillary gemmae and rhizoid tubers lacking; not known west of the Mississippi River ..... 2. BRACHYMENIUM
  2. Plants terrestrial or on logs and stumps (sometimes on tree bases); leaves mostly not twisted around the stems; axillary gemmae and rhizoid tubers common; distribution various 3
3. Leaves often bordered with narrow, elongate cells; leaf cells mostly short, ± rhombic-hexagonal; costa mostly ending in leaf apex or excurrent (sometimes ending below apex) ..... 4. BRYUM
3. Leaves unbordered; leaf cells linear; costa mostly ending in or below leaf apex ..... 1. POHLIA

Plants small to medium-size; stems erect, simple or forked; leaves narrow, lanceolate, not conspicuously bordered; margins plane or recurved; cells narrow, elongate, often linear; costa mostly ending

1. POHLIA Hedw.

in or below leaf tip. Setae elongate; capsules mostly inclined to pendulous, neck short or long; operculum convex to short-rostrate.

The genus *Pohlia* is similar in many respects to *Bryum*, differing mainly in the elongate leaf cells and the unbordered leaves. *Pohlia* is well represented to the north but rare along the Gulf Coast. We have two species.

Plants bearing abundant tubers on rhizoids and often in axils of
    lower leaves; stem tips fragile, easily detached   1. P. APICULATA
Plants lacking tubers; stem tips not fragile, remaining attached   2. P. WAHLENBERGII

1. Pohlia apiculata (Schwaegr.) Crum & Anderson
Figure 35, E–G

Bryum cruegeri Hampe in C.M.

Plants small, green to yellowish green, often with reddish tinge, ± glossy, gregarious or in soft sods; rhizoids reddish brown, bearing abundant reddish brown to dark purple tubers with protuberant cells, to ca 200 μm diameter, the tubers often present in lower leaf axils as well; stems erect, red below, rather fleshy and fragile, the tips detaching readily, to several cm long but mostly much shorter, mostly simple; leaves often distant and spreading on lower part of stem, imbricate above, not contorted when dry, reddish at base, only slightly concave, oblong-lanceolate to narrowly lanceolate, acute to broadly acute; margins plane, entire below, sometimes irregularly serrate at apex, marginal cells long and narrow but scarcely forming a distinct border; cells elongate-rhombic to fusiform, quadrate at basal angles; costa often red or brown, ending in apex to shortly excurrent, usually decurrent. Not producing sporophytes in our area.

> **Habitat:** Roadsides, flower beds, greenhouses, along streams; soil and rock.
> **Range:** Pantropical.
> **Gulf Coast Distribution:** Mississippi, Louisiana, Texas.

This moss is easily recognized by the fleshy, often fragile stems, ± flat, uncontorted leaves, and the axillary and rhizoidal tubers with bulging cells. In vigorous colonies the stem tips break away freely when brushed gently, and it is likely that they serve as propagula. Recent studies by Ochi (1974) and others have shown that this moss has a worldwide distribution in the warmer regions. However, it is apparently quite rare along the Gulf Coast. I have seen authentic specimens from Louisiana, Mississippi, and Texas in our area. It may occur in great abundance in greenhouse pots and benches. *Pohlia apiculata* is better known in the Americas under the names *Bryum cruegeri* and *Pohlia cruegeri*.

2. Pohlia wahlenbergii (Web. & Mohr) Andr.
Figure 35, H–I

Plants pale, whitish glaucous when fresh, in ± soft tufts or thin, loose clumps; stems red (at least at base), erect to inclined, simple or forked; leaves erect-appressed to spreading when dry, imbricate to distant, lanceolate to narrowly lanceolate or triangular, acute;

cells oblong-rhomboidal to linear; margins plane or somewhat recurved, serrate toward apex; costa strong, ending near or in leaf
apex or shortly excurrent into a sharp, narrow point. Not producing
sporophytes in our area.

Habitat: Ravine banks in rich mesic forests; stream banks and
seepage areas.
Range: Widespread in the Northern Hemisphere.
Gulf Coast Distribution: Louisiana

This inconspicuous moss is very rare on the Gulf Coast, where it
is known only from a few collections in Louisiana. The plants
somewhat resemble a coarse *Philonotis* in general aspect. It was
originally reported from Louisiana as *P. pulchella* (Hedw.) Lindb.
The reddish stems, soft texture, and glaucous appearance, coupled
with the elongate cells, make it rather easy to recognize. It was
reported from Harrison County, Texas, by Whitehouse and McAllister (1954), but I have not seen the specimen.

POHLIA NUTANS (Hedw.) Lindb. has been reported from Mobile
County, Alabama (Wilkes 1955), but I have not been able to study
the specimen. It seems unlikely that this moss actually occurs on
the Gulf Coast.

Plants scattered to loosely tufted, dark to reddish or brownish
green; stems fleshy, erect, mostly simple, densely matted below
with dark red to brownish tomentum, short, flagelliform branchlets
often present in upper leaf axils of older stems; leaves twisted
around stem when dry, erect-spreading when moist, oblong to
ovate-lanceolate, decurrent, mostly bordered in upper part with
narrow, thick-walled, brownish cells, border mostly lacking below
midleaf, sometimes lacking altogether; cells lax, mostly broadly
elongate-rhombic, square to rectangular in leaf base; leaf margin
serrulate at apex, recurved; costa strong, excurrent into a stout,
concolorous, sometimes toothed awn. Setae elongate; capsules
erect, elliptic, contracted at both ends, with short, inconspicuous
neck; peristome short, pale, segments shorter than teeth, both very
papillose; spores smooth to finely papillose; operculum bluntly
conic, slightly oblique.

Habitat: On tree bark in mesic to moist forests; along streams
and rivers; usually on *Magnolia grandiflora* and live oak;
often growing with *Schlotheimia rugifolia*.
Range: Southeastern United States; Mexico; Guatemala; South
America.
Gulf Coast Distribution: Florida to southeastern Louisiana.

This little moss is quite rare along the Gulf Coast. It is easy to
recognize because of its habitat and the leaves twisted around the
stems when dry. In other parts of its range it also grows on rock and
soil, but in our area it has only been found on trees.

2. BRACHYMENIUM Schwaegr.

Brachymenium klotzschii
(Schwaegr.) Par.
Figure 36, A–B

Brachymenium macrocarpum
Card.

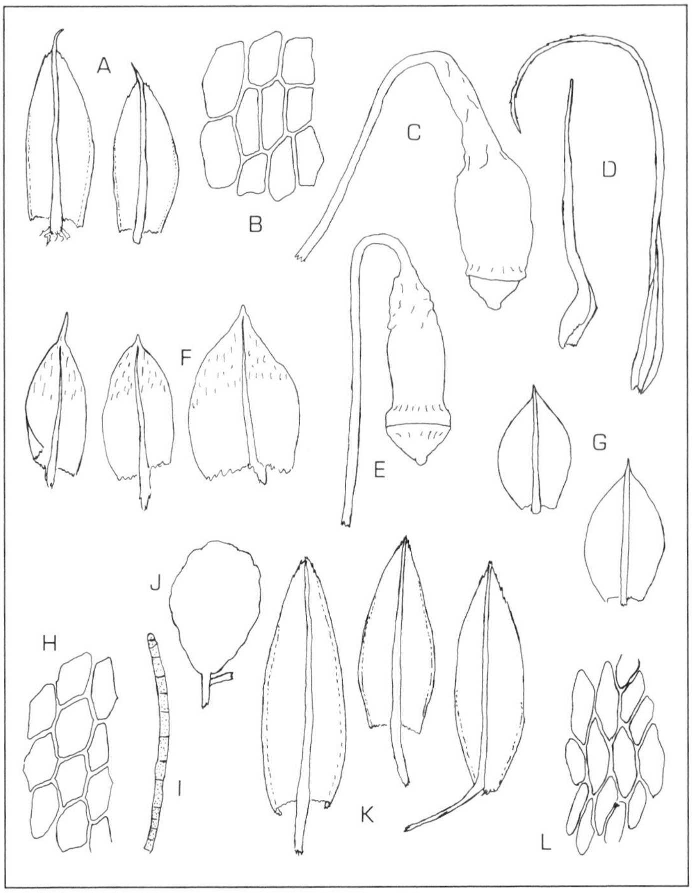

Fig. 36. *A–B*: **Brachymenium klotschii.** *A*, leaves (× 15). *B*, upper median cells (× 231). *C–D*: **Leptobryum pyriforme.** *C*, capsule (× 15). *D*, leaves (× 11). *E–F*: **Bryum argenteum.** *E*, capsule (× 15). *F*, leaves (× 33). *G–J*: **Bryum capillare.** *G*, leaves (× 15). *H*, median cells (× 231). *I*, gemma (× 55). *J*, tuber (× 55). *K–L*: **Bryum pseudotriquetrum.** *K*, leaves (× 15). *L*, median cells (× 231).

3. LEPTOBRYUM (B.S.G.) Wils.

Leptobryum pyriforme
(Hedw.) Wils.
Figure 36, C–D

Plants slender, delicate, in ± glossy tufts or clumps, sometimes soft and silky; stems erect, dark purple to blackish below; rhizoids pale to dark purple, sometimes brownish, papillose, sometimes bearing smooth, purple or brownish, multicellular gemmae; leaves remote and reduced below, crowded and elongate at stem tips; upper leaves flexuous when dry, subulate, to 6 mm long, composed mostly of the costa above the somewhat expanded base; margins entire below, serrate-toothed toward apex; costa long-excurrent in upper leaves. Setae elongate, flexuous, reddish; capsules inclined to pendulous, with prominent neck as long as urn; operculum convex to umbonate; spores smooth to papillose.

> **Habitat:** Soil, humus, damp brickwork; frequent in greenhouses.
> **Range:** Essentially cosmopolitan.
> **Gulf Coast Distribution:** Louisiana.

This moss is apparently rare or absent over most of the Gulf Coast, where it seems to be known only from a few collections from brick structures in Louisiana. The plants are somewhat reminiscent of *Ditrichum,* but the dark purple stem bases, purple rhizoids, and the sporophyte distinguish it from plants of that genus. The gemmae occasionally occur in the lower leaf axils as well as on the rhizoids. Faith Pennebaker Mackaness collected large quantities of this moss from tombs in old cemeteries in New Orleans; the specimens are in the herbarium of Tulane University, New Orleans.

4. BRYUM Hedw.

Plants small to medium-size or robust, yellowish green to green or darker, often with pink to red (or darker) tinge, gregarious to tufted or in dense turves; rhizoid tubers present or lacking, sometimes in axils of lower leaves; axillary bulbils or filamentous gemmae produced by some; stems erect, short or elongate, usually forked, loosely to densely foliated, often matted with tomentum below; leaves mostly remote and reduced below, larger and crowded at stem tips, mostly variously lanceolate to triangular and acute to acuminate, not much contorted when dry; upper cells mostly rhombic, sometimes narrow and longer, basal cells broader and more lax; margins plane to recurved, entire to serrate, often ± bordered with elongate cells; costa ending below leaf apex to strongly excurrent. Setae elongate; capsules mostly inclined to pendulous; peristome double; operculum convex to conic, mostly ± umbonate.

    Bryum is a large and taxonomically difficult genus; Howard Crum (1976) aptly referred to it as "the *Crataegus* of bryology." Specific characteristics in *Bryum* reside in large part in the sexual condition and in the sporophyte; unfortunately, most *Bryum* species on the Gulf Coast produce sporophytes only rarely—or not at all. Therefore, I have used characteristics of the gametophore exclusively in the following key, with the realization that such a key may be difficult to use in some cases and will not always render a satisfactory identification.

Special Study Methods for *Bryum*

    It is usually easy to find the rhizoid tubers, if present, by breaking apart the sod of moss plants and searching for the tubers, under a dissecting microscope, on the freshly exposed soil-rhizoid fracture surface. The tubers are usually glossy, and often dark, making them fairly easy to spot in the soil-rhizoid matrix. They are sometimes very abundant and conspicuous.

1. Plants small, silvery; leaves hyaline in upper half	1. B. ARGENTEUM
1. Plants small to medium-size or larger, various shades of green but not silvery; leaves green throughout	2
2. Plants bearing rhizoid tubers, or filamentous or spherical axillary gemmae or bulbils, or both	3
2. Plants lacking tubers, gemmae, and bulbils	5
3. Axillary gemmae filamentous; rhizoid tubers commonly present; leaves ± twisted around stem	2. B. CAPILLARE

3. Axillary gemmae (if present) spherical, or plants bearing bulbils only; rhizoid tubers present or lacking 4

    4. Bulbils present in upper leaf axils (sometimes scarce) at least on older, larger stems; rhizoid tubers lacking .... 5. B. BICOLOR

    4. Bulbils lacking; rhizoid tubers always present .... 6. B. RADICULOSUM

5. Costa percurrent or slightly excurrent; plants relatively robust .... 3. B. PSEUDOTRIQUETRUM

5. Costa mostly strongly excurrent; plants small to medium-size .... 4. B. LISAE var. CUSPIDATUM

---

1. **Bryum argenteum** Hedw.
Figure 36, E–F

Plants small, silvery white, gregarious or in low tufts or sods; stems short, branched, ± erect, julaceous, red, naked below; leaves very concave, not contorted when dry, imbricate, orbicular to oblong or elliptic, abruptly short-acuminate, green to reddish in lower part, hyaline above; cells rather thick-walled, fusiform to rectangular or short-linear, quadrate in leaf base; costa usually excurrent into a stout hyaline awn, sometimes ending below leaf apex. Setae yellow to red; capsules yellow to red, pendulous, short-cylindric to oblong, neck short; spores smooth; operculum short-conic, with small, blunt, oblique apiculation.

> **Habitat:** Disturbed sites in ± open areas; paths, old fields, sidewalks, parking lots.
> **Range:** Cosmopolitan.
> **Gulf Coast Distribution:** Florida to Texas.

> *Bryum argenteum* is one of the easiest of all mosses to recognize because of its silvery aspect and habitat. It is fairly common, even weedy, throughout our area but is not often conspicuous. Sporophytes are not commonly produced on the Gulf Coast but are sometimes found in abundance. *Bryum argenteum* is one of the few mosses in commercial trade. It is used, for example, in bonsai plantings, under the name "silver moss."

2. **Bryum capillare** Hedw.
Figure 36, G–J

Plants small, dark green, ± glossy, gregarious, tufted, or in low, dense sods, often with dark red tinge, mostly matted with dark reddish brown rhizoids below, almost always with abundant, filamentous, uniseriate, finely papillose gemmae clustered in the leaf axils and usually with abundant glossy tubers on the rhizoids, the tubers red when young, purple-black when older (as seen in reflected light); stems erect, short, simple or forked; leaves not usually much contorted, often somewhat twisted around the stem when dry, crowded to separated, ovate to lanceolate, short-acuminate, margins usually weakly bordered with 1–3 rows of narrow, elongate cells, especially above, entire except at the often toothed apex; upper cells firm, hexagonal to rhombic, longer than wide, quadrate in leaf base; costa ending well below apex to conspicuously excurrent into a stout concolorous awn. Rarely producing sporophytes on the Gulf Coast; setae reddish; capsules inclined to pendulous, ± cylindric, with prominent neck, to *ca* 4 mm long including the neck; operculum umbonate.

> **Habitat:** Growing on rock, soil, humus, stumps, logs, tree bases; low forests, swamps, along streams; especially common in wet forests and swamps.
> **Range:** Europe; North America, including Mexico.
> **Gulf Coast Distribution:** Florida to Texas.

Bryum capillare is rather common along the Gulf Coast, almost weedy. It is easy to recognize by the usually twisted leaves and the filamentous axillary gemmae, which are almost always present. Occasional specimens lack the gemmae, or the rhizoid tubers, or both; however, both are usually present. The gemmae are often quite conspicuous and can be seen easily with a hand lens when abundant. *Bryum flaccidum* Brid. [*B. capillare* var. *flaccidum* (Brid.) B.S.G.] may be a better name for this species, based on Syed's (1973) publication, in which he points out that, in the strict sense, *B. capillare* lacks axillary gemmae.

Plants medium-size to ± robust, dark green to reddish, in dense tufts or clumps; rhizoid tubers lacking; stems conspicuously red, mostly simple, matted with brown tomentum below; leaves shriveled-contorted when dry and sometimes ± twisted around stem, often well separated below, lanceolate, often red across base; cells firm, hexagonal, slightly elongate, rectangular in leaf base; margins revolute, conspicuously bordered by elongate thick-walled cells (border sometimes difficult to see because of revolute margins), denticulate above or entire; costa strong, percurrent, red at base and decurrent. Not producing sporophytes in our area.

3. Bryum pseudotriquetrum
(Hedw.) Gaertn., Meyer & Scherb.
Figure 36, K–L

> **Habitat:** Creek banks; roadsides in wet forests.
> **Range:** Circumpolar; North and South America; Australia.
> **Gulf Coast Distribution:** Florida to eastern Texas.

This is a very distinctive moss that is easy to recognize by its stature, red stem with matted tomentum below, percurrent costa that is red below and decurrent, and revolute leaf margins. It lacks the axillary filamentous gemmae of *B. capillare* and the excurrent costa of *B. lisae* var. *cuspidatum*. Although present over much of the Gulf Coast, it is nowhere common or abundant here.

Plants dark green to yellowish, medium-size, in dense clumps or tufts; rhizoid tubers lacking; stems short, red below, with reddish brown tomentum; leaves crowded at stem tips, somewhat shriveled and twisted when dry, lanceolate-acuminate, about 3 mm long, often reddish at base; cells oblong-rhombic, basal cells larger, ± rectangular; margins revolute, at least below, distinctly bordered by narrow, elongated, thick-walled cells; costa long-excurrent into a sharp, stout, toothed or entire awn. Synoicous; sporophytes rare in

4. Bryum lisae De Not. var.
cuspidatum (B.S.G.) Marg.
Figure 37, A–B

Bryum cuspidatum (B.S.G.)
Schimp.

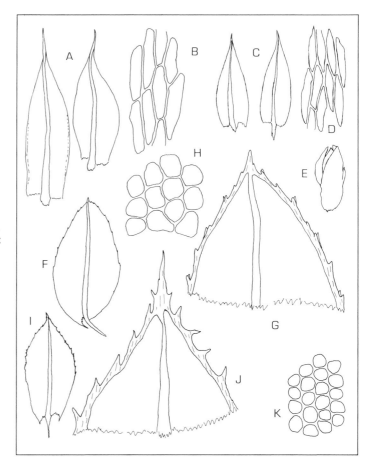

Fig. 37. *A–B*: **Bryum lisae** var. **cuspidatum.** *A*, leaves (× 15). *B*, median cells (× 231). *C–E*: **Bryum bicolor.** *C*, leaves (× 26). *D*, median cells (× 231). *E*, bulbil (× 55). *F–H*: **Mnium thomsonii.** *F*, leaf (× 15). *G*, leaf tip (× 55). *H*, median cells (× 231). *I–K*: **Mnium cuspidatum.** *I*, leaf (× 9). *J*, leaf tip (× 55). *K*, median cells (× 231).

our area; capsules inclined to ± pendulous, *ca* 3 mm long, with rather short neck.

Habitat: Rock and soil; old fields, roadsides.
Range: North America; circumpolar.
Gulf Coast Distribution: Louisiana; Mississippi.

This moss is rare in our area; I have seen only a few specimens, mostly from Louisiana. The long-excurrent costa, lack of rhizoid tubers, and distinctly bordered leaves make it fairly distinctive in our area. This *Bryum* is somewhat similar at first glance to *B. pseudotriquetrum*, but that species is more robust, has much longer stems with the leaves more evenly spaced, and the costa is only percurrent (or sometimes very shortly excurrent).

5. **Bryum bicolor** Dicks.
Figure 37, C–E

Plants small, in dense, soft, green to dark green sods, rhizoid tubers lacking or very scarce (not seen in Gulf Coast specimens); stems erect, slender, mostly simple, to 15 mm long but mostly much shorter, at least the older, larger stems usually bearing abundant (sometimes scarce), leafy bulbils in upper leaf axils; leaves not contorted when dry, closely imbricate, concave, ovate, short-acuminate, about 1 mm long; cells rhomboidal, basal cells quadrate to rectangular; margins essentially unbordered, plane to ± recurved, especially at base; costa strong, often brown or reddish, ending in apex or excurrent into a stout awn. Not producing sporophytes in our area.

Habitat: Roadsides, cracks in pavements and walls, around buildings, tree bases, rock and soil; often in disturbed sites.
Range: Essentially cosmopolitan.
Gulf Coast Distribution: Florida to Texas.

Smith and Whitehouse (1978) recognized four species in the *B. bicolor* complex in Britain, basing their conclusions in large part on the nature of the bulbils. It is likely that close study of American species in this complex would reveal a similar situation. Until such a study is performed, however, it seems best to use the name *B. bicolor* in its traditional sense and to not attempt to recognize additional taxa. The bulbils, which make this moss easy to recognize, are apparently nearly always present, although sometimes scarce and only on larger, older stems. Also helpful in recognition is the absence of rhizoid tubers, which are usually present in abundance on some of the other troublesome little species of *Bryum* found on disturbed sites along the Gulf Coast. The bulbils can usually be easily detected by holding the moss sod upside down over a piece of white paper and gently brushing the stem tips with your hand. The bulbils will dehisce and fall on the paper.

6. **Bryum radiculosum** Brid.

Plants mostly rather small and slender; yellowish green to green, scattered, gregarious, or in dense low turves; rhizoidal tubers always present and often very abundant; stems mostly reddish below, very short to ca 15 mm long; leaves ovate-lanceolate to lanceolate or triangular, imbricate, not or little-contorted when dry; cells elongate-hexagonal, shorter and wider at leaf base; margins often recurved, entire or denticulate above, with or without distinct border; costa strong, mostly excurrent, sometimes strongly so. Rarely producing sporophytes in our area; capsules pendulous, short, ± ovoid, with prominent neck.

This is a troublesome group of weedy little mosses. They are not often collected, perhaps because the colonies are usually inconspicuous and sterile and often grow in sites that are unattractive to most collectors, such as trampled areas (as in parks and pathways), roadsides, parking lots, and around buildings in urban areas. The rhizoid tubers are always present and often produced in great abundance. With a little practice it is easy to find the tubers in the soil below the moss mat, as well as among the decaying stem bases in the lower part of the mat itself. Because the species often grow intermixed with one another, and with other mosses which may also bear tubers, care must be taken to associate the recovered tubers with the proper moss.

These mosses are poorly known along the Gulf Coast, and over most of the rest of the world as well. At our present state of knowledge, it is not possible to identify satisfactorily every specimen encountered of these small tuber-bearing plants. For further information on tuber-bearing mosses see Whitehouse (1966); for treatments of the European and British taxa, respectively, of this complex, see Crundwell and Nyholm (1964) and Smith (1978). See Crum and Anderson (1981) for information on the American taxa. Preliminary studies on Gulf Coast specimens indicate that perhaps as many as six different segregates of *B. radiculosum* occur here. However, it is not possible at this time to assign formal taxonomic names to them.

BRYUM CAESPITICIUM Hedw. has been reported from our area, in Alabama and Texas. It is very similar to *B. lisae* var. *cuspidatum* but differs in being dioicous and in its more slenderly acuminate leaves with longer, narrower upper cells, and its usually very short stem (the plants often bulbiform). It is perhaps not possible to make a positive identification of sterile material.

BRYUM CORONATUM Schwaegr. has been reported from our area, in Mississippi. However, I have not been able to locate the specimen. Crum and Anderson (1981) attribute this pantropical moss only to southern Florida in the U.S.

BRYUM TURBINATUM (Hedw.) Turn. was reported from Burleson County, Texas, by Magill (1973). Crum and Anderson (1981), however, express doubt that this species actually occurs in North America.

MNIACEAE Schwaegr.

Plants small to rather robust, green to dark green or sometimes reddish, in thin to dense mats or tufts, sometimes forming extensive carpets; sterile, prostrate, flattened, creeping stems often present with the leaves appearing 2-ranked; sexual stems erect; leaves increasing in size upward, obovate to elliptic or oblong-elliptic, pointed, mostly bordered with elongate, narrow cells and mostly toothed on the margins; cells mostly rounded-quadrate or hexagonal and with firm walls, smooth, becoming rectangular below; costa single, mostly ending just below the leaf apex. Sporophytes terminal; setae elongate; capsules usually inclined to pendulous and symmetric; peristome double; operculum hemispheric to apiculate or rostrate; calyptra mostly cucullate.

The Mniaceae are mainly woodland mosses of often moist habitats. Our species are all treated under the genus *Mnium*, although the genus is sometimes subdivided into several genera.

The name *Mnium* is familiar to many people who have had a college course in botany because *Mnium* is frequently used as an example of the mosses in textbooks and in laboratory courses. Our species grow on soil, especially on stream banks, and on rocks, logs, and stumps. Individual colonies may form conspicuous mats of vegetation in favorable habitats and are especially noticeable in the winter and early spring, while the forest canopy is open. *Mnium* plants are tough and fairly hardy; the large leaves on the creeping or ascending, flattened sterile stems of some species give the plants an attractive aspect, like a miniature fern. The qualities of durability and pleasing appearance make some species of *Mnium* popular as terrarium plants.

Although *Mnium* resembles some species of *Bryum* to a certain extent, the two should not be confused. *Bryum* lacks the sterile, creeping, flattened stems often found in *Mnium*, and plants of *Bryum* are mostly softer than those of *Mnium*. The leaf cells of *Bryum* are usually more delicate in appearance and usually more elongate than in *Mnium*.

Special Study Methods for *Mnium*

Dry plants of *Mnium* soak up cold water very slowly. Prior to dissection for study, dip the dried plants into hot water for a moment or two to soften them. It may be desirable to determine the sexual condition. If so, look for the erect, sexual stems in the specimen. The sexual stems are usually easy to find because of the enlarged leaves surrounding the inflorescence at the stem tip. Carefully dissect out the sex organs in a drop of water on a slide and examine the preparation to see if only antheridia or archegonia are present (dioicous), or if both types of sex organs are present in the same inflorescence (synoicous). A similar search may be made in the perichaetial leaves surrounding the base of a seta, but the sex organs in a mature perichaetium supporting a sporophyte are usually shriveled or sometimes virtually absent.

MNIUM Hedw.

1. Leaf margins bearing paired teeth	1. M. THOMSONII
1. Leaf margins with the teeth single (teeth sometimes much reduced or nearly lacking)	2
2. Leaves toothed only in the upper half	2. M. CUSPIDATUM
2. Leaves toothed nearly to the base	3. M. AFFINE

Plants in thin to dense, reddish green mats; sterile stems often highly complanate and with the leaves appearing to be 2-ranked; sexual stems erect; leaves ovate, erect to erect-spreading and contorted when dry, erect-spreading when moist, bordered all around

1. **Mnium thomsonii** Schimp. Figure 37, F–H

Mnium orthorhynchum, (in the sense of Breen, 1963).

with elongate cells; margins toothed mainly in the upper half, the teeth mostly in pairs; leaf cells isodiametric; costa excurrent in short apiculation. Dioicous. Sporophytes usually one per perichaetium; capsules inclined to horizontal; operculum rostrate.

> **Habitat:** Rocks, logs, soil, bark at tree bases, in rich forests.
> **Range:** Northern Hemisphere.
> **Gulf Coast Distribution:** Known in our area only from Florida Caverns State Park, Jackson County, Florida.

Its reddish green color and doubly-toothed leaf margins distinguish *M. thomsonii* easily from our other species of *Mnium*. It is possible that this species will turn up in other sites west of Florida. Prime habitats to search include calcareous exposures in rich forests. This moss was long known in North America under the name *M. orthorhynchum*.

2. Mnium cuspidatum Hedw.
Figure 37, I–K

Plants in dark to yellowish green mats; sterile stems ± complanate and prostrate; sexual stems erect; leaves erect-contorted when dry, erect-spreading when moist, obovate, acute to broadly acuminate, bordered all around with elongate cells; margins strongly toothed in upper half; leaf cells isodiametric, thick-walled, thickened at the corners, in rows; costa ending in the leaf apex in a sharp apiculation. Synoicous. Sporophytes one per perichaetium; capsules pendulous; operculum hemispheric.

> **Habitat:** On soil, logs, tree bases in rich, well-drained forests; common on ravine banks.
> **Range:** Northern Hemisphere.
> **Gulf Coast Distribution:** Florida to Texas.

This species is not nearly so common or abundant as *M. affine* but is nonetheless widely distributed throughout our area. It occurs in upland habitats not subject to flooding, in contrast to *M. affine* which often grows in low-lying habitats where water accumulates. In forested ravines, *M. cuspidatum* is commonly found on the ravine banks whereas *M. affine* occurs along the bottom of the ravine. The darker color of *M. cuspidatum* and its erect or erect-spreading leaves distinguish it well from *M. affine* macroscopically. Under the microscope the obovate leaves, toothed only in the upper half, distinguish *M. cuspidatum* from *M. affine*, which has elliptical leaves usually toothed all around. When fertile stems are present, the synoicous condition of *M. cuspidatum* will aid in distinguishing it from the dioicous *M. affine*.

3. Mnium affine Bland.
ex Funck
Figure 38, A–C

Plants in thin to dense, green to yellowish green mats; sterile stems ± complanate, sometimes attenuated; sexual stems erect; stems usually with abundant light brown rhizoids; leaves contorted-recurved when dry, spreading when moist, broadly elliptic to oblong, bordered all around with elongate cells; leaf margins

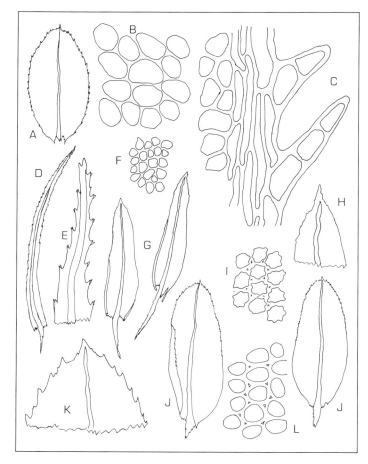

Fig. 38. *A–C*: **Mnium affine.** *A*, leaf (× 7). *B*, median cells (× 231). *C*, margin in upper part of leaf (× 231). *D–F*: **Rhizogonium spiniforme.** *D*, leaf (× 11). *E*, leaf tip (× 55). *F*, median cells (× 231). *G–I*: **Aulacomnium palustre.** *G*, leaves (× 15). *H*, leaf tip (× 55). *I*, median cells (× 436). *J–L*: **Aulacomnium heterostichum.** *J*, leaves (× 15). *K*, leaf tip (× 55). *L*, median cells (× 436).

toothed with teeth of 1–3 cells, the teeth sometimes reduced or nearly absent; leaf cells ± isodiametric or slightly elongate, often in oblique rows, thick-walled; costa slightly excurrent, ending in a sharp apiculus. Dioicous. Sporophytes usually one per perichaetium; capsules pendulous; operculum hemispheric to apiculate.

> **Habitat:** Stream banks, ravines; moist, shady places in lawns; on soil, logs, tree bases.
> **Range:** Northern Hemisphere; South America.
> **Gulf Coast Distribution:** Florida to eastern Texas.

Mnium affine is common and often abundant along much of the Gulf Coast; it is the only species of *Mnium* present in most of our area. In rich, upland sites, *M. cuspidatum* may also be present, in better-drained habitats than *M. affine*. In *M. affine* the leaves are toothed only in the upper half, and the plants are synoicous. The spreading to recurved leaves of *M. affine*, even when dry, also aid in distinguishing it from *M. cuspidatum*, in which the leaves are erect to erect-spreading. In the latter species too, the leaves are less contorted when dry than are those of *M. affine*. Generally, the two species can be distinguished in the field. The leaves in some colonies of *M. affine* bear much-reduced marginal teeth.

MNIUM LONGIROSTRUM Brid. (M. ROSTRATUM Schrad.) has been reported a few times from the Gulf Coast, but I have not seen any specimens of it from our area. Its synoicous sexual condition, oblong leaves toothed all around with inconspicuous, single-celled teeth, and rostrate operculum distinguish it from *M. affine*, which it most closely resembles of our species of *Mnium*.

RHIZOGONIACEAE Broth.

Plants robust; stems erect; leaves narrow, toothed on margins; cells smooth, isodiametric, thick-walled; costa single, strong, ending in leaf apex. Sporophytes basal; setae elongate; capsules curved, inclined to horizontal; peristome double; operculum obliquely conic-rostrate; calyptra cylindric, smooth, cucullate, fimbriate at base.

This mainly tropical family is represented in our area only by a single species, in the genus *Rhizogonium*.

RHIZOGONIUM Brid.

Rhizogonium spiniforme
(Hedw.) Bruch ex Kraus
Figure 38, D–F

Plants robust, dark to yellowish or brownish green, in soft, loose, springy cushions or tufts; stems erect, simple, to about 8 cm tall, sometimes curved at the tips; leaves flexuous, not much contorted when dry, erect-spreading when moist, decurrent, linear-acuminate, tapering gradually to the narrow, acute, toothed apex; cells rounded to square, thick-walled; leaf margins slightly thickened, prominently toothed with sharp, paired teeth in upper ¾ or nearly to base; costa strongly toothed in upper portion, percurrent. Fairly commonly fruiting in our area; sporophytes as described for the family.

> **Habitat:** Humid forests and deep, wooded ravines; on soil, stumps, roots, logs, in wet places, especially around springs and seeps.
> **Range:** Tropical and subtropical regions around the world.
> **Gulf Coast Distribution:** Florida to southeastern Louisiana.

This is a graceful and beautiful moss. It is sometimes found in great abundance in especially favorable sites where the lush, green mounds that it forms are very conspicuous. At first glance, *R. spiniforme* could be taken for a *Mnium*; however, its leaves are not bordered by elongate cells as they are in our species of *Mnium*, and the sporophytes arise basally, rather than terminally as in *Mnium*. It is not really common over most of our area and has not yet been found west of the Mississippi River.

AULACOMNIACEAE Schimp.

Plants moderately robust, green to yellowish green or brownish, mostly in clumps or sods but sometimes gregarious, often with abundant brown to reddish brown rhizoids in lower portions; stems ± erect, often branched, sometimes ± flattened, frequently with naked, setalike apical extensions which may bear small, leaflike brood bodies; leaves narrow to broad, toothed (in our species); cells smooth to unipapillose, ± isodiametric; costa single, strong, ending in or near leaf apex. Sporophytes terminal; setae elongate; capsules inclined to horizontal, somewhat curved, striate when dry; peristome double; operculum conic to rostrate; calyptra cucullate, smooth.

This family is represented in our area by one genus, *Aulacomnium*, with two species.

The species of *Aulacomnium* are primarily northern in their range. Our two species are forest mosses, one growing in rather wet sites and the other in dry habitats. The latter, *A. heterostichum*, is fairly common over much of our area, but the other species, *A. palustre*, is uncommon. The leaves of both species have evident intercellular spaces at the corners of the cells, a quality that is lacking in most mosses.

AULACOMNIUM Schwaegr.

Leaves narrow, acute, entire or toothed only near the tips, not homomallous; stems not complanate	1. A. PALUSTRE
Leaves broad, rounded-apiculate above, toothed to the middle, homomallous; stems usually complanate	2. A. HETEROSTICHUM

Plants yellowish green, in loose to dense clumps or scattered, mostly 1–4 cm tall in our area; leaves somewhat contorted when dry, erect-spreading when moist, decurrent, narrowly ovate-lanceolate to linear-lanceolate, apex acute; margins recurved below or nearly to apex, entire or ± toothed near apex; cells isodiametric above, basal ones elongate, cell walls thickened, sinuous; each cell with a low, often inconspicuous papilla; costa sinuous above, ending near or in leaf apex. Not fruiting in our area.

1. Aulacomnium palustre (Hedw.) Schwaegr. Figure 38, G–I

> **Habitat:** Soil, tree bases, roots, and rocks, in wet, forested areas.
> **Range:** Nearly worldwide; across North America.
> **Gulf Coast Distribution:** Florida to Texas.

This species is widespread in our area but rather rare. The yellowish green color and naked, often propaguliferous extensions of the stems make it easy to recognize in the field, where it almost always occurs in rather wet sites. Under the microscope the re-curved leaf margins, papillose cells with sinuous walls, and the toothed or irregular margins just below the leaf apex are characteristic. The papillae on the leaf cells are sometimes very inconspicuous. More or less conspicuous intercellular spaces are sometimes present between leaf cells, especially at the corners of the cells. See remarks under *A. heterostichum*.

2. **Aulacomnium heterostichum** (Hedw.) B.S.G.
Figure 38, J–L

Plants green to yellowish green, in dense clumps; stems usually complanate, branched, 1–2 cm tall; leaves decurrent, homomallous, somewhat rugose but not much contorted otherwise when dry, ovate, often folded on one side in lower part, apex rounded-apiculate, margins conspicuously toothed in upper ⅓–½, somewhat recurved on one or both sides at base; cells rounded, thick-walled, smooth or inconspicuously unipapillose, basal ones scarcely differentiated; costa strong, straight or slightly sinuous, ending in or near leaf tip. Infrequently fruiting in our area; capsules inclined, striate when dry; operculum obliquely rostrate.

> **Habitat:** Well-drained banks and tree bases in forests.
> **Range:** Asia; eastern North America; Mexico.
> **Gulf Coast Distribution:** Florida to Texas.

This attractive little moss is fairly common in well-drained forested areas along the Gulf Coast. Its commonest habitat is right on the lips of ravines, where there is an abrupt change from the level forest floor to the slope of the ravine side. The naked, propaguliferous stem extensions of this species are not as conspicuous as are those of *A. palustre*. The two species are not really similar in habit, habitat, or appearance, and thus are easy to tell apart. *Aulacomnium heterostichum* might be confused at first sight with *Mnium*, but it lacks the elongate marginal cells present in our species of that genus. Beginning students of mosses have taken *A. heterostichum* for a *Fissidens*, at first glance, because of its often rather flattened stems. Intercellular spaces, at the cell corners, are also present in the leaves of this species, but are not as obvious as in *A. palustre*.

BARTRAMIACEAE Schwaegr. in Willd.

Small to fairly robust plants in dense or thin, pale green to yellowish green or glaucous clumps; stems erect or lax, mostly not much branched, often matted with reddish brown rhizoids in older portions; leaves ovate-lanceolate to linear-lanceolate or triangular, straight or flexuous when dry, erect-spreading when moist; cells thick-walled, quadrate to elongate, mostly unipapillose, the papillae at the upper ends of the cells; margins toothed, often revolute in part or nearly throughout, uni- or bistratose; costa single, strong, ending well below the leaf tip to strongly excurrent. Sporophytes terminal but usually quickly overtopped by innovations and then appearing lateral; setae elongate; capsules inclined to horizontal, usually furrowed; operculum convex or conic; peristome double; calyptra small, cucullate, naked. Rarely fruiting in our area.

Plants of the Bartramiaceae grow mostly in moist to damp or wet sites, often along streams and on seeping banks and rocks. They can often be recognized in the field by the glaucous or yellowish green color and by the habitat. The family is represented in our area by the genus *Philonotis*.

Philonotis is very common in our area and occurs essentially throughout the Gulf Coast. Plants of *Philonotis* prefer damp sites in general, often in disturbed areas such as eroding stream and ravine banks. The pale, glaucous or yellowish green color of the erect, usually short stems is characteristic. The treatment used here is based on that of Zales (1973), according to which three species occur in our area. Plants of *Philonotis* only rarely produce sporophytes along the Gulf Coast but often bear abundant deciduous branchlets that serve for vegetative reproduction and dispersal.

PHILONOTIS Brid.

Special Study Methods for *Philonotis*

Plants of *Philonotis* are extremely slow to soak up water for dissection, even when fresh from the field. Immersing the plants briefly in boiling water will soften them and allow their dissection. Zales (1973) recommends clearing the leaves of *Philonotis* in lactophenol in order to see the papillae and other details more clearly; this is not really necessary, however, in identifying our Gulf Coast species.

1. Leaf cells quadrate to 4 times as long as wide; leaves ovate-triangular to triangular, apex acute or blunt; costa ending in or below leaf apex — 1. P. GLAUCESCENS
1. Leaf cells 6–20 times as long as wide; leaves triangular, apex acute; costa excurrent — 2
    2. Leaves long and narrowly triangular; cells long and narrow, to 20 times as long as wide, papillae bluntly pointed, at the extreme upper ends of the cells; monoicous — 2. P. LONGISETA
    2. Leaves triangular to slightly ovate-lanceolate; cells rectangular, less than 9 times as long as wide, papillae bluntly rounded, near or at the upper ends of the cells; dioicous — 3. P. MARCHICA

Plants slender, to 2 cm tall but usually shorter, in light green to yellowish green, thin colonies; stems erect or lax, not much branched; leaves erect-spreading, ovate-oblong to triangular-lanceolate, apex acute to bluntly rounded; cells quadrate to rhombic or short-rectangular, essentially smooth or each with a blunt

1. **Philonotis glaucescens** (Hornsch.) Broth. Figure 39, A–B

Philonotis gracillima Ångstr.

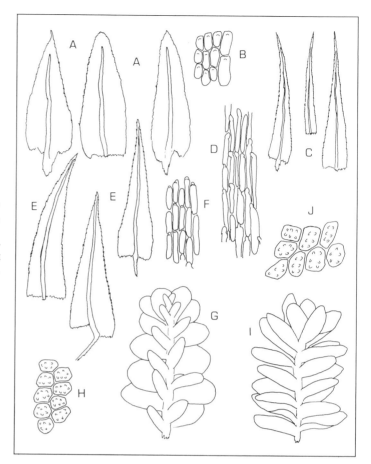

Fig. 39. *A–B:* **Philonotis glaucescens.** *A,* leaves (× 55). *B,* median cells (× 247). *C–D:* **Philonotis longiseta.** *C,* leaves (× 21). *D,* median cells (× 247). *E–F:* **Philonotis marchica.** *E,* leaves (× 27). *F,* median cells (× 247). *G–H:* **Erpodium biseriatum.** *G,* habit, ventral view (× 25). *H,* median cells (× 189). *I–J:* **Erpodium domingense.** *I,* habit, ventral view. *J,* median cells (× 189).

papilla at the upper end on the adaxial surface; leaf margins serrate; costa ending below leaf tip to percurrent. Probably not fruiting in our area.

Habitat: On soil and rock in moist, protected sites.
Range: Southeastern United States; Mexico; Central America; West Indies; South America.
Gulf Coast Distribution: Florida, Mississippi, Louisiana, Texas.

This species is rather rare in our area. It has not yet been reported from Alabama but surely occurs there. See comments under *P. longiseta.*

2. Philonotis longiseta
(Michx.) E. G. Britt.
Figure 39, C–D

Plants slender, to *ca* 3 cm tall but mostly much shorter, green to yellowish green, in dense clumps or thin colonies; stems erect, little branched; leaves sometimes ± secund, long and narrowly triangular to lanceolate-acuminate, apex acute; cells long and narrow, in vertical rows, most cells with a stout, bluntly pointed papilla at the upper end on the adaxial surface; leaf margins serrate, sometimes

± revolute; costa excurrent, forming the leaf tip. Rarely fruiting in our area.

Habitat: Soil or rock in wet sites; ravine banks, ditches, gullies, road cuts, along streams.
Range: Southeastern North America; Mexico; Central America; Puerto Rico.
Gulf Coast Distribution: Florida to Texas.

This is the most common and abundant Gulf Coast species of *Philonotis*; the two other species that occur in our area are much less likely to be encountered. The narrowly triangular leaves, excurrent costa, and linear cells with prominent papillae distinguish *P. longiseta* from *P. glaucescens*; and its longer and narrower cells, stouter and more pointed papillae, and more slenderly pointed leaves distinguish it from *P. marchica*.

3. **Philonotis marchica** (Hedw.) Brid.
Figure 39, E–F

Plants slender, to *ca* 6 cm tall but mostly shorter, green to yellowish green, in soft mats or tufts, densely covered with brown rhizoids in lower portions; stems little-branched, erect or lax; leaves erect-spreading, triangular to ± ovate-lanceolate, acuminate, apex acute; cells linear-oblong, most cells with a bluntly rounded papilla at the upper end; leaf margins serrate, sometimes revolute; costa excurrent, usually forming the leaf tip. Probably not fruiting in our area.

Habitat: Wet soil on stream banks; in water.
Range: North America; Europe; Mexico.
Gulf Coast Distribution: Louisiana, Texas.

Philonotis marchica is rare in our area, where it is known from a single station in Louisiana and from several in Texas; it has not yet been reported from other areas along the Gulf Coast. See comments under *P. longiseta*. *Philonotis marchica* apparently grows in wetter sites than does *P. longiseta*.

Philonotis sphaerocarpa (Hedw.) Brid., P. uncinata (Schwaegr.) Brid. and P. gracillima Ångstr. have been reported from the Gulf Coast in our area. However, according to Zales (1973), the first two names represent but a single species (*P. sphaerocarpa*) that does not occur in our area, and the third name, *P. gracillima*, is treated by Zales as a synonym of *P. glaucescens*.

Bartramia pomiformis Hedw. has been reported from our area in Mississippi. I have not been able to see the specimen upon which the report was based and doubt that this species actually occurs so far south. It is easily distinguished from *Philonotis* by its leaves 4–6 mm long and slenderly long-acuminate from a sheathing base.

ERPODIACEAE Broth.

Plants small, green to yellowish or brownish, in thin mats; stems prostrate, creeping, irregularly branched; leaves often complanate, sometimes asymmetric, occasionally 4-ranked, all alike or dimorphic, oblong to elliptic, apex rounded to acute or acuminate; cells ± isodiametric to a little elongate, smooth or pluripapillose; margins entire; costa lacking. Sporophytes terminal on short branches; setae short; capsules immersed to shortly exserted, pale, symmetric, oblong-cylindric; operculum apiculate to rostrate; peristome none or rudimentary; calyptra mitrate or cucullate, small.

The Erpodiaceae are widespread, mainly in the tropical and subtropical parts of the world. The family is represented on the Gulf Coast by a single genus, *Erpodium*.

ERPODIUM (Brid.) C.M.

Two species of *Erpodium* occur along the Gulf Coast; both have very restricted distributions in our area and one of them, *E. biseriatum*, is very rare. They are very small, inconspicuous mosses and are not easy to detect in the field.

Leaves dimorphic, the dorsal leaves much larger than the ventral leaves; plants strongly flattened when dry; not known west of Louisiana on the Gulf Coast .... 1. E. BISERIATUM

Leaves all alike; plants somewhat julaceous or only slightly flattened when dry; known in our area only from the lower Rio Grande valley .... 2. E. DOMINGENSE

1. **Erpodium biseriatum** (Aust.) Aust.
Figure 39, G–H

Solmsiella kurzii Steere

Plants very small, dark green to yellowish or brownish green, in thin, flat mats, very much with the aspect of a small, leafy liverwort; stems prostrate, flattened, with short branches; dorsal leaves spreading, in two rows, elliptic to oblong-ovate, rounded at apex; ventral leaves usually much smaller than dorsal ones, spreading at about 45°, elliptic to lingulate, obtuse; cells isodiametric or slightly elongate, pluripapillose, with rather thick walls. Sporophytes borne on short branches; capsules slightly exserted, cylindric; operculum obliquely apiculate; calyptra small, ± cucullate.

Habitat: Bark of large trees, especially *Magnolia grandiflora*, in rich, well-drained forests.

Range: Southeastern United States; Mexico; South America; southeastern Asia; India; Ceylon; Java.

Gulf Coast Distribution: Florida and Louisiana.

Erpodium biseriatum is known from only a very few collections in the United States—in Georgia, Florida, and Louisiana. It could very well be more common and widespread along the Gulf Coast than the few collections indicate, but the small size of the plants and their remarkably close resemblance to leafy liverworts make it difficult to find the plants in the field. The liverwortlike aspect of *E. biseriatum* is reflected in the fact that the plants were first described as a species of *Lejeunea*, a genus of leafy liverworts. Under the microscope, the pluripapillose cells, leaves in four rows, and lack of lobules, underleaves, and oil bodies all point to its being a moss rather than a liverwort. The presence of sporophytes would, of course, identify it as a moss immediately.

The prime habitat in which to seek *E. biseriatum* is the smooth bark of large trees in rich, humid forests. Perhaps the best method would be to collect in the field all plants resembling leafy liverworts and then to search for the *Erpodium* under the dissecting microscope.

Plants dark green to brownish green, in thin, intricate mats; stems prostrate, creeping, ± turgid to slightly flattened when dry, somewhat flattened when moist; leaves all alike, spreading when moist, oblong to elliptic, obtuse, in four rows; cells rather obscure, isodiametric, densely pluripapillose, somewhat thick-walled. Sporophytes on short branches; capsules shortly exserted, oblong; operculum obliquely rostrate; calyptra very small, narrow, mitrate.

2. Erpodium domingense
(Spreng.) C.M.
Figure 39, I–J

> **Habitat:** Tree bark, rotted wood, rock; in dry, often rather open forests and scrub areas.
> **Range:** Texas; Mexico; Central America; West Indies; Ecuador.
> **Gulf Coast Distribution:** The lower Rio Grande valley of Texas.

Erpodium domingense is known in Texas from several collections in Cameron County and from one in Hidalgo County. It is probably not rare in wooded areas in the lower Rio Grande valley, although wooded areas there are themselves rare.

PTYCHOMITRIACEAE Schimp.

Plants small, in low, dark green to blackish or brownish cushions or tufts; stems erect to decumbent, usually freely forked; leaves ± contorted to coiled, or straight, when dry, spreading to erect-spreading when moist, narrowly to broadly acuminate from a broader, somewhat sheathing base, bistratose above, at least on the margins, sometimes throughout the upper part; cells ± isodiametric or slightly transversely elongate, obscure, thick-walled, smooth to bulging or slightly papillose, sometimes a little porose, cells in basal part of leaf pellucid, elongate; margins entire to irregular or coarsely toothed; costa strong, single, percurrent or nearly so. Sporophytes terminal; setae short; capsules ovoid to elliptic; operculum slenderly long-rostrate; peristome single; calyptra campanulate-mitrate, plicate.

These dark, low-growing mosses are most closely related to the Grimmiaceae. One genus, *Ptychomitrium*, occurs in the Gulf coastal plain.

PTYCHOMITRIUM Fürnr.   Three species of *Ptychomitrium* are known from the Gulf coastal plain; two of the species are common and abundant, the third is very rare in our area.

---

1. Leaves 3–3.5 mm long, coarsely toothed above with large teeth; plants growing primarily on calcareous rocks .... 3. P. SERRATUM
1. Leaves 1.5–2.5 mm long, entire or irregularly slightly serrate above; plants growing on trees or logs, or on siliceous rocks 2
   2. Leaves erect, straight or slightly contorted when dry, often somewhat serrate above; plants growing on trees or logs, dull .... 1. P. DRUMMONDII
   2. Leaves curled when dry, entire above; plants growing on siliceous rocks, glossy .... 2. P. INCURVUM

---

1. Ptychomitrium drummondii (Wils.) Sull.
Figure 40, A–C

Plants small, in low, dense, dark green to blackish colonies; leaves dull, ± erect and straight or somewhat contorted when dry, spreading when moist, 1.5–2 mm long, apex acute, rather narrowly acuminate from a broader base, unistratose in base, uni- to bistratose in upper part; cells in upper part of leaf opaque, thick-walled, often transversely elongate, smooth or bulging, sometimes slightly papillose; cells of leaf base large, pellucid, rectangular; margins unistratose below, bistratose above, entire to irregular or somewhat serrate above; costa percurrent. Usually fruiting freely in our area; sporophytes yellowish; capsules ovoid, ribbed when dry; peristome teeth recurved when dry; calyptra campanulate-mitrate, lobed at base.

> **Habitat:** Tree trunks and bases, logs; in humid areas, frequent along rivers but also in urban areas on trees along streets and on old roofs. A few collections have been taken from rock or soil.
> **Range:** Eastern United States.
> **Gulf Coast Distribution:** Florida to Texas.

Ptychomitrium drummondii is very common and abundant along much of the Gulf Coast; however, the colonies are often not very conspicuous except when they are fruiting. The abundantly produced, yellowish sporophytes make the colonies easy to detect when they are in fruit. The dull aspect and only slightly contorted leaves, when dry, distinguish *P. drummondii* in the field from *P. incurvum*, as does the habitat. *Ptychomitrium drummondii* often

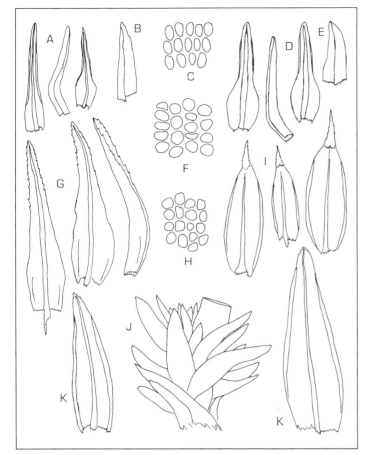

Fig. 40. *A–C*: **Ptychomitrium drummondii.** *A*, leaves (× 14). *B*, leaf apex (× 55). *C*, median cells (× 436). *D–F*: **Ptychomitrium incurvum.** *D*, leaves (× 14). *E*, leaf apex (× 55). *F*, median cells (× 412). *G–H*: **Ptychomitrium serratum.** *G*, leaves (× 15). *H*, median cells (× 412). *I*, **Orthotrichum diaphanum,** leaves (× 15). *J–K*: **Orthotrichum pusillum.** *J*, habit, moist (× 15). *K*, leaves (× 26).

forms very large colonies in favorable sites, such as large, leaning trees along bayous and lakes.

Plants small, in low, dense, blackish green sods; leaves glossy, tightly contorted-curled when dry, spreading when moist, 1.5–2.5 mm long, apex acute to blunt, cucullate, narrowly or broadly acuminate from a broader base, bistratose above, cells as in *P. drummondii* except flat; margins entire; costa disappearing below leaf tip. Operculum slenderly conic-rostrate; calyptra campanulate.

Habitat: More or less shaded sandstone rocks and cliffs; in dry, rather open forests.

Range: Eastern North America; Europe.

Gulf Coast Distribution: Florida, Louisiana, and Texas.

Ptychomitrium incurvum is pretty much restricted to siliceous rocks; thus its occurrence on the Gulf coastal plain is determined by the occurrence of such substrates. It was collected once in southern Louisiana, where no siliceous rocks occur naturally, on sandstone boulders which had been imported for landscaping pur-

2. **Ptychomitrium incurvum**
(Schwaegr.) Spruce
Figure 40, D–F

poses. This species has not yet been reported from southern Alabama or Mississippi. At first sight sterile plants of this little moss could be mistaken for a *Weissia*, but the leaves are not involute as in that genus. In fertile specimens, the campanulate calyptra, as opposed to the cucullate calyptra of *Weissia*, is distinctive. See comments under *P. drummondii*.

3. **Ptychomitrium serratum**
(C.M.) B.S.G. ex Besch.
Figure 40, G–H

Plants medium-size, in small, dull, brownish green to blackish tufts; leaves contorted when dry, erect-spreading when moist, 3–3.5 mm long, broadly to narrowly acuminate from a somewhat sheathing base, the base often plicate, apex acute, leaves uni- or bistratose above; cells smooth, quadrate to rounded, ± porose above; margins coarsely toothed in upper part with large teeth; costa percurrent. Sporophytes larger than those of the two preceding species; capsules ovoid-cylindric, ribbed; operculum rostrate; calyptra campanulate, covering the capsule.

> **Habitat:** Calcareous rocks, including concrete.
> **Range:** Louisiana; Texas; Mexico; Guatemala; South America.
> **Gulf Coast Distribution:** Louisiana and Texas.

Ptychomitrium serratum is larger in all respects than either *P. drummondii* or *P. incurvum*. Its stature, habitat, and coarsely toothed leaves make it easy to distinguish. This moss is almost entirely restricted to calcareous substrates. Three of the four Gulf Coast collections of this very rare (in our area) moss were taken from concrete structures: a dam spillway and a discarded block of concrete in Louisiana, and a picnic area fireplace in Texas. The fourth collection was taken from a 17-year-old asphalt shingle roof in southern Louisiana. One of the Louisiana collections was fruiting; the other Gulf Coast collections were sterile. It is interesting to speculate on how this species arrived in eastern Texas and southern Louisiana, where it occurs on man-made substrates, from its only other known station in the U.S., the Guadeloupe Mountains of Culberson County, Texas.

ORTHOTRICHACEAE Arnott

Plants small, green to blackish or dark reddish green, in small tufts or low mats; stems erect and mostly simple (*Orthotrichum*) or prostrate and with short, erect branches (*Groutiella*, *Macromitrium*, *Schlotheimia*); leaves straight and erect-appressed when dry, or spirally twisted around the stems, or contorted, erect to spreading when moist, short-lanceolate to lingulate; upper cells small, isodiametric, smooth to bulging or papillose, basal cells mostly elongate; margins entire or slightly toothed above, sometimes revolute; costa strong, single, percurrent to excurrent. Sporophytes terminal on stems and branches; setae short to elongate; capsules ovoid to short-cylindric, immersed to exserted, ribbed or smooth when dry and empty; operculum short- to long-rostrate; peristome single or double, or lacking; calyptra cucullate, campanulate, or mitrate, often hairy.

This is a rather large family of mostly corticolous mosses. Four genera occur along the Gulf Coast, with best representation in the eastern part.

1. Stems erect; plants forming small clumps; capsules ± immersed on very short setae	1. ORTHOTRICHUM
1. Stems prostrate, with short, erect branches; plants forming mats; capsules exserted on elongate setae	2
2. Leaves spirally twisted around the stem when dry; cells smooth or bulging	3
2. Leaves crispate-contorted when dry, but not spirally twisted around the stem; cells papillose	2. MACROMITRIUM
3. Leaves strongly rugose both wet and dry, without border of elongate cells	4. SCHLOTHEIMIA
3. Leaves not rugose, bordered in lower ⅓ to ½ with elongate cells	3. GROUTIELLA

Plants small, erect, green to dark green; leaves straight and tightly appressed when dry, ovate-lanceolate to oblong, apex obtuse, acute, or acuminate; leaf margins revolute, entire, or somewhat denticulate in upper part; cells smooth or papillose. Setae very short; capsules ovate-oblong to cylindric, smooth to wrinkled or ribbed when dry; operculum conic to short-rostrate; peristome single or double; calyptra oblong-campanulate, smooth to slightly hairy.

Orthotrichum is a large genus, with about forty species in North America, of which only two species occur in our area along the Gulf Coast. One species, *O. pusillum*, is fairly common in our area, although rarely collected; the other species is infrequent on the Gulf Coast. Our species of *Orthotrichum* grow almost exclusively on the bark of hardwood trees. The small size and inconspicuous nature of *Orthotrichum* plants make them difficult to find in the field except by a special search; *Orthotrichum* plants are not likely to be collected by the casual observer of mosses.

1. ORTHOTRICHUM
Hedw.

Leaf tips hyaline	1. O. DIAPHANUM
Leaf tips concolorous, not hyaline	2. O. PUSILLUM

Plants small, green to dark green, in small clumps of a few plants; leaf tips (at least on the upper leaves) elongated into a conspicuously hyaline and usually serrate awn. Capsules immersed to emergent, pale, smooth to somewhat ribbed.

1. Orthotrichum diaphanum
Brid.
Figure 40, I

Habitat: Tree bark in dry, often more or less open sites.

Range: Widespread in the Northern Hemisphere; western Tennessee to Utah and Arizona, and south to Louisiana and Mexico.

Gulf Coast Distribution: Northern Louisiana, Texas.

Orthotrichum diaphanum is easy to recognize in the field because the hyaline leaf tips are readily seen with a hand lens; the same quality distinguishes *O. diaphanum* from our other species of *Orthotrichum*. The plants are small and often grow intermixed with other species of *Orthotrichum*, thus making them difficult to find.

2. **Orthotrichum pusillum**
Mitt.
Figure 40, J–K

Plants small, green to blackish green or brownish, growing in small clusters of a few plants or forming more extensive colonies; leaf tips bluntly acute, sometimes denticulate; upper cells of leaves mostly 10–13 μm diameter. Capsules immersed to emergent, pale, smooth to wrinkled or slightly ribbed when dry, not particularly contracted below the mouth when empty.

Habitat: Bark of hardwood trees in rather open sites.

Range: Eastern United States.

Gulf Coast Distribution: Florida to eastern Texas.

This little moss is rather common in our area but is not often collected due to its small size and inconspicuous nature. Pecan, black walnut, and mulberry are among the trees that it favors as a habitat, but it has been collected from many others too, including elm, oak, hackberry, red gum, and magnolia.

ORTHOTRICHUM OHIOENSE Sull. & Lesq. and O. PUMILUM Sw. have been reported from along the Gulf Coast. However, the distributions of these two species are mainly more northerly, and they probably do not occur in our range. Both are similar to *O. pusillum* but have darker capsules, usually contracted below the mouth. The capsule is strongly ribbed in *O. pumilum*, which has the upper leaf cells mostly 14–20 μm diameter; the capsule is less strongly ribbed in *O. ohioense*, whose upper leaf cells are mostly 8–10 μm diameter. The pale, smooth or only slightly ribbed capsule of *O. pusillum*, not evidently contracted below the mouth, distinguishes it from the other two. I have not seen any convincing specimens of either *O. pumilum* or *O. ohioense* from our range.

2. **MACROMITRIUM Brid.**

Macromitrium richardii
Schwaegr.
Figure 41, A–B

Plants small, green to reddish green, in thin or dense low mats; primary stems prostrate and creeping, becoming densely brown-tomentose and bearing reduced leaves; secondary stems erect, little-branched; leaves strongly crispate-contorted when dry, spreading when moist, ovate- to linear-lanceolate, apex bluntly acute; upper cells quadrate to rounded, papillose, lower cells vertically elongate, with thickened, sinuose-pitted walls; margins plane above, ± recurved on one side below; costa ending below leaf tip.

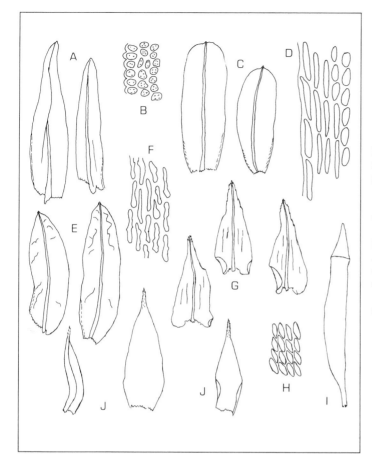

Fig. 41. *A–B:* **Macromitrium richardii.** *A,* leaves (× 26). *B,* margin and cells near leaf tip (× 379). *C–D:* **Groutiella tumidula.** *C,* leaves (× 26). *D,* margin and adjacent cells near leaf base (× 379). *E–F:* **Schlotheimia rugifolia.** *E,* leaves (× 26). *F,* basal cells (× 379). *G–I:* **Climacium americanum.** *G,* leaves (× 15). *H,* median cells (× 231). *I,* capsule (× 15). *J,* **Hedwigia ciliata,** leaves (× 15).

Sporophyte terminal on branches; setae elongate; capsules ovoid, strongly constricted at mouth when dry, ribbed; calyptra campanulate, fringed at base, bearing scanty straight hairs in upper part; operculum slenderly rostrate; peristome single.

Habitat: Bark of hardwood trees in rather open, mesic to wet forest.

Range: Southeastern United States; tropical America.

Gulf Coast Distribution: Florida; Mississippi; southeastern Louisiana.

Although this species somewhat resembles *Schlotheimia rugifolia* at first glance, its nonrugose, highly contorted (when dry) leaves, which are not spirally twisted about the stem, will distinguish it easily under a hand lens. *Macromitrium richardii* is rare in our area. It has not yet been reported from Alabama but surely occurs there. The plants are less robust than those of *Schlotheimia rugifolia* and are generally much less conspicuous in the field.

3. GROUTIELLA Steere

Groutiella tumidula (Mitt.) Vitt
Figure 41, C–D

Groutiella mucronifolia (Hook. & Grev.) Crum & Steere (in the sense of Breen, 1963).

Plants medium-size, green to brownish or reddish, in dense, intricate mats; primary stems prostrate, creeping, thickly matted with reddish brown tomentum; secondary stems ascending, elongate, little-branched; leaves ± spirally twisted around stems, ovate-lanceolate to broadly linear, mucronate to shortly acuminate, bordered in lower ⅓–½ with elongate cells; upper cells rounded, thick-walled, bulging, a few basal cells elongate and with thickened, sinuous-pitted walls; margins plane above, ± recurved on one side below; costa excurrent into the acumination. Sporophytes terminal on branches; setae elongate; capsules short-cylindric; operculum slenderly rostrate; peristome lacking; calyptra campanulate, fringed at base, naked.

> **Habitat:** Bark of tree trunks and branches.
> **Range:** Florida; tropical America.
> **Gulf Coast Distribution:** Florida.

This is a very distinctive species that is known in the range of this manual only from Jackson County, Florida. It could well turn up, however, farther west along the coast. It is fairly frequent in southern Florida. Although the leaves are spirally twisted around the stems when dry, similar to those of *Schlotheimia rugifolia*, they are not rugose as in the latter species. The usually well-defined border on the lower portion of the leaves is also distinctive.

4. SCHLOTHEIMIA Brid.

Schlotheimia rugifolia (Hook.) Schwaegr.
Figure 41, E–F

Plants medium-size, reddish to brownish or dark green, forming low, dense mats; primary stems prostrate, creeping, densely brown-tomentose, bearing reduced leaves concealed within the tomentum; secondary stems erect, crowded, simple or sparingly branched; leaves ± spirally twisted around the stems when dry, distinctly rugose both wet and dry, ovate-oblong, mucronate; cells smooth, upper cells quadrate to rounded, thick-walled, lower cells vertically elongate, with narrow lumens and greatly thickened, sinuose-pitted walls; costa shortly excurrent and forming the mucro; margins plane above, ± revolute on one side below. Sporophyte terminal on the secondary stems or appearing lateral; setae elongate; capsules cylindric, erect; peristome double; operculum long-rostrate; calyptra rough above, naked, campanulate-mitrate, fringed at base.

> **Habitat:** Bark of trees in mesic to wet forests; occasionally on rock, humus, soil, and decaying logs and stumps.
> **Range:** Southeastern United States; tropical America.
> **Gulf Coast Distribution:** Florida to eastern Texas.

Schlotheimia rugifolia is a rather attractive moss when growing in its typical mat form on tree trunks. The often reddish color of the plants is distinctive and permits recognition of *Schlotheimia* in the field from some distance away. This moss superficially resembles in habit two of its much rarer (in our area) relatives, *Macromitrium richardii* and *Groutiella tumidula*. However, neither of the latter two species has the rugose leaves of *S. rugifolia*, which are easily seen with a hand lens. While the leaves of the *Groutiella* are more or less spirally twisted around the stem when dry, as in *Schlotheimia*, those of *Macromitrium* are crispate but not twisted.

Schlotheimia rugifolia is fairly common and often abundant along the Gulf Coast, especially from Louisiana to the east.

FONTINALACEAE Schimp.

Plants small and slender to rather robust, lax to rigid, dark, growing in water or at least immersed at high water levels, forming thin colonies or large, dense, intricate masses of highly branched, elongate stems; leaves in three rows (but this is not usually evident except in *Dichelyma*), ± erect to falcate-secund, ovate to lanceolate, bluntly acute to long and slenderly acuminate; leaf cells shortly elongate to linear, smooth; costa lacking or single and well developed and percurrent to long-excurrent. Sporophytes terminal on short branches; setae very short to elongate; capsule immersed to exserted; operculum conic to rostrate; peristome double, segments of endostome often united by lateral projections and forming a "trellis"; calyptra conic to mitrate or cucullate.

The Fontinalaceae are represented in our area by three genera. *Fontinalis*, with five taxa along the Gulf Coast, is commonly encountered if one seeks in the appropriate habitats; *Brachelyma*, with one species in our area, is infrequent; and *Dichelyma*, also with a single species in our area, is rare. Plants of the Fontinalaceae grow attached to various substrates in standing or flowing water, although they are commonly collected at times when the water level has receded and the colonies of plants are exposed. An intriguing question is how plants of the Fontinalaceae, which are usually sterile along the Gulf Coast, achieve upstream distributions. I have often found colonies of *Fontinalis* in small, temporary pools in gullies on rather steep hillsides, where water only flows following rain and where there are long reaches of dry gully between pools. How does *Fontinalis* become established in these isolated situations?

Like many aquatic plants, members of the Fontinalaceae serve as a habitat for myriads of other forms of aquatic life—diatoms and other algae and insect larvae, for example. As a consequence, plants of this family are sometimes coated with scum when taken from the water and with a crust of debris when dried. Sometimes only the young stem and branch tips will be free of such biological encrustation. In dried specimens, the glassy shells of diatoms are often very conspicuous under the dissecting microscope.

1. Leaves ecostate	1. FONTINALIS
1. Leaves with well-developed costa	2
2. Costa percurrent or ending just below leaf apex	2. BRACHELYMA
2. Costa very long-excurrent	3. DICHELYMA

1. FONTINALIS Hedw.

Plants mostly dark to yellowish green, slender to robust, lax to ± rigid, forming dense, tangled growths; stems freely branched; leaves erect to somewhat spreading, ovate to lanceolate, alar cells ± differentiated; costa lacking. Apparently rarely producing sporophytes in our area.

According to Howard Crum (1976), *Fontinalis* is "surely the most difficult genus of mosses." Those who attempt to key out specimens of *Fontinalis*, in this text or any other, will very likely have every reason to agree with him. Fourteen species of *Fontinalis* are known in eastern North America (Crum and Anderson 1981), and five of them occur along the Gulf Coast. One (*F. langloisii*) is restricted to southeastern Louisiana and adjacent Mississippi; the others have wider distributions in our area.

Colonies of *Fontinalis* often form huge floating masses in flowing streams or quiet pools and swamps. The plants are often collected, however, after water levels have fallen, leaving the mats of *Fontinalis* dry and suspended conspicuously from tree bases and roots.

Special Study Methods for *Fontinalis*

The following key to *Fontinalis* is based on median stem leaves, that is, on leaves taken from a well-developed stem some distance below its tip. Avoid using leaves from branches or slender, poorly developed stems.

Identification of *Fontinalis* is often difficult and fraught with uncertainty. The robust *F. novae-angliae* is fairly easy to recognize, as are the soft *F. flaccida*, the very slender *F. filiformis*, and the more or less rigid *F. sullivantii*. However, it may be difficult to separate *F. langloisii* from *F. filiformis*. Aquatic plants in general are notoriously variable in form and appearance. Changes in growth form due to seasonality, rise and fall of water levels, and turbidity, for example, all may have subtle or profound effects on the appearance of aquatic plants.

1. Leaves mostly plane	1. F. FLACCIDA
1. Leaves mostly concave	2
2. Plants very slender, leafy stems and branches narrow	3
2. Plants medium-size to robust, leafy stems and branches thicker	4
3. Plants lax; leaves to about 0.5 mm wide, tips acuminate	2. F. FILIFORMIS
3. Plants firm to rigid; leaves to about 1.4 mm wide, tips often blunt	3. F. LANGLOISII

4. Branch leaves clearly smaller than stem leaves; plants rigid ..... 4. F. SULLIVANTII

4. Branch leaves similar to stem leaves, intergrading in size; plants soft, not particularly rigid ..... 5. F. NOVAE-ANGLIAE

Plants medium-size, lax to ± rigid, glossy; stems highly branched; leaves plane, soft, narrowly lanceolate to ovate-lanceolate, 5–7 mm long, 0.5–1.5 mm wide; alar cells enlarged; auricles conspicuous; leaf tips long-acuminate.

1. Fontinalis flaccida Ren. & Card.
Figure 42, A–B

> **Habitat:** Attached to various substrates in rivers, streams, and swamps; mostly in flowing water.
> **Range:** Eastern North America.
> **Gulf Coast Distribution:** Alabama and Louisiana.

Although this *Fontinalis* has apparently not been reported along the Gulf Coast from the Florida Panhandle, Mississippi, or Texas, it surely occurs there. It is fairly common in Louisiana and has been collected in several parishes along the Sabine River, adjacent to Texas.

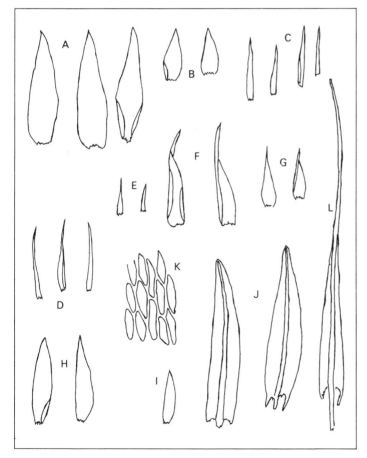

Fig. 42. *A–B:* **Fontinalis flaccida.** *A,* stem leaves (× 5). *B,* branch leaves (× 5). *C,* **Fontinalis filiformis,** stem leaves (× 5). *D–E:* **Fontinalis langloisii.** *D,* stem leaves (× 5). *E,* branch leaves (× 5). *F–G:* **Fontinalis sullivantii.** *F,* stem leaves (× 5). *G,* branch leaves (× 5). *H–I:* **Fontinalis novae-angliae.** *H,* stem leaves (× 5). *I,* branch leaf (× 5). *J–K:* **Brachelyma subulatum.** *J,* leaves (× 16). *K,* median cells (× 328). *L,* **Dichelyma capillaceum,** leaf (× 16).

2. Fontinalis filiformis Sull. &
Lesq. ex Aust.
Figure 42, C

Plants very slender and lax, usually greatly elongated; leaves concave to tubulose, very narrowly lanceolate, 3–6 mm long, 0.3–0.5
mm wide; alar cells slightly enlarged; auricles ± distinct; leaf tips
subulate-acuminate.

> **Habitat:** Attached to logs, trees, shrubs, in rivers, streams, and
> sloughs.
> **Range:** Eastern United States.
> **Gulf Coast Distribution:** Florida to eastern Texas.

This species has not yet been reported from the Florida Panhandle but surely occurs there. It is, perhaps, the easiest of all of
our species of *Fontinalis* to recognize, due to the much elongated,
extremely slender and lax stems. It often grows in huge, floating
masses. *Fontinalis langloisii* is similar but differs evidently in its
quite rigid habit, especially noticeable when the plants are dry. In
contrast, plants of *F. filiformis* are soft and lax both wet and dry.

3. Fontinalis langloisii Card.
Figure 42, D–E

Plants slender, ± rigid; leaves plane to concave or tubular, narrowly lanceolate, 3.5–5.5 mm long, 0.9–1.4 mm wide, alar cells
± enlarged, auricles absent or present but inconspicuous; leaf tips
obtuse to ± acute.

> **Habitat:** Streams, ponds; on trees and logs.
> **Range:** Southern United States.
> **Gulf Coast Distribution:** Louisiana and Mississippi.

Fontinalis langloisii is not a well-known moss and has apparently not been collected in recent years. Its total known distribution
is in St. Tammany Parish, Louisiana, and adjacent Hancock County,
Mississippi. See comments under *F. filiformis*.

4. Fontinalis sullivantii Lindb.
Figure 42, F–G

Fontinalis disticha Hook. &
Wils.

Plants rather robust, ± rigid; leaves concave, at least at base, narrowly lanceolate, 4–6.5 mm long, 0.7–1.5 mm wide; alar cells
enlarged, auricles ± distinct; leaf tips narrowly acuminate.

> **Habitat:** Attached to various substrates in rivers, swamps,
> streams, and sloughs.
> **Range:** Eastern United States.
> **Gulf Coast Distribution:** Florida to eastern Texas.

This species is known under the name *F. disticha* Hook. & Wils.
in most books and herbaria. It is fairly distinct in its ± rigid habit
and in the leaves concave at least in the lower portion, although
often plane above. The long, narrowly lanceolate leaves with acuminate tips also help to identify it. *Fontinalis filiformis* is much
more slender in all its parts and much more lax too.

5. Fontinalis novae-angliae
Sull.
Figure 42, H–I

Plants medium to robust, often ± rigid; leaves concave, ovate- to
oblong-lanceolate or lanceolate, one or both margins involute, *ca*
2.5–4 mm long, *ca* 0.7–2 mm wide, alar cells enlarged and conspicuous, forming distinct auricles; leaf tips mostly obtuse.

Habitat: Attached to roots, tree bases, rocks, etc., in flowing or still waters; rivers, streams, sloughs.
Range: Eastern North America.
Gulf Coast Distribution: Florida to eastern Texas.

Its rather robust nature and leaves intergrading in size between stems and branches make this species fairly easy to recognize. Some forms are quite large and may produce large floating mats in appropriate habitats.

FONTINALIS HYPNOIDES C. J. Hartm. has been reported from our area, from Polk County, Texas. I have not seen the specimen and doubt that this species actually occurs in our area. Its leaves are plane, like those of *F. flaccida*, but shorter (only 3–4.5 mm long) and with blunter apices.

2. BRACHELYMA Schimp. ex Card.

Brachelyma subulatum
(P.-Beauv.) Schimp. ex Card.
Figure 42, J–K

Brachelyma robustum (Card.) E. G. Britt.

Plants slender to robust, lax to ± rigid, dark green to yellowish or brownish, in intricate tangles; stems highly branched, elongate; leaves erect to somewhat spreading, oblong-lanceolate, ± bordered with elongate cells, occasionally somewhat secund; costa percurrent or ending just below leaf apex. Not producing sporophytes in our area.

Habitat: Attached to tree bases, shrubs, roots, logs, twigs, pilings, etc., in rivers, sloughs, and swamps.
Range: Southeastern United States.
Gulf Coast Distribution: Florida to eastern Texas.

Brachelyma subulatum is only infrequently collected along the Gulf Coast, but often occurs in great abundance when it is found. It is more common to the east along the Gulf than to the west; it is known in Texas only from Hardin and Harrison counties. Plants of *B. subulatum* vary a good deal in robustness; some are very slender and lax, others more robust and rigid; the latter have been recognized taxonomically as *B. robustum* (Card.) E. G. Britt., but this name is generally considered now to be a synonym of *B. subulatum*. The robust form is much less frequent than the slender form and is sometimes strikingly different in aspect; it has not been collected west of Florida. Slender, lax forms of *B. subulatum* are sometimes very much like *Fontinalis* in appearance; however, the presence of a costa is diagnostic.

3. DICHELYMA Myr.

Dichelyma capillaceum
(With.) Myr.
Figure 42, L

Plants slender, short or elongated, yellow-brown or darker, in thin or dense colonies; stems irregularly branched; leaves usually ± secund, especially at stem tips, linear-lanceolate, very slenderly acuminate; costa long-excurrent, forming the acumination. Not producing sporophytes on the Gulf Coast.

Habitat: Sloughs, river swamps; on tree bases, shrubs, and exposed roots.

Range: Europe, eastern North America.
Gulf Coast Distribution: Louisiana, Mississippi.

Dichelyma capillaceum is rare along the Gulf Coast. It is known only from Tangipahoa and Vernon parishes, and also from New Orleans, in Louisiana, and from Amite County in Mississippi. It has not been reported from the Florida Panhandle, Alabama, or Texas. The Louisiana station for *D. capillaceum* in Vernon Parish is in the drainage of the Sabine River, which forms the boundary between Louisiana and Texas, so the moss could be expected to occur in adjacent regions of Texas. *Dichelyma capillaceum* is easily recognized in the field because of its habitat and the long, slenderly pointed leaves. At first glance one unfamiliar with it might mistake *D. capillaceum* for a slender dicranaceous moss, due to the secund leaves. Look for it in sloughs and swamps along rivers and bayous, in summer and fall after water levels have fallen.

CLIMACIACEAE Kindb.

Plants robust; primary stems prostrate, secondary stems erect and ± dendroid; paraphyllia abundant; leaves oblong-ovate to ovate-lanceolate, auriculate, toothed at least in upper part, ± plicate; costa single, strong, ending in acumination; cells rhomboidal to linear, smooth. Dioicous; setae elongate, capsules cylindric, erect and symmetric in our species; peristome double; operculum rostrate; calyptra cucullate. We have only one species, in the genus *Climacium*, along the Gulf Coast.

CLIMACIUM Web. & Mohr
Climacium americanum Brid.
Figure 41, G–I

Plants robust, tough, dull to dark green; primary stem prostrate, creeping on substrate or subterranean, bearing reduced leaves and abundant brownish rhizoids; secondary stems erect, dendroid (sometimes ± simple in plants from flooded sites); filiform, branched or simple paraphyllia abundant on secondary stems and branches; leaves broadly ovate-lanceolate to broadly triangular, ± plicate, auriculate, coarsely toothed above, teeth smaller below, apex blunt; cells rhomboidal to shortly elongated, smooth; costa ending in acumination, often ± sinuous. Sporophytes infrequently produced, borne on stems and bases of branches in upper part of plant, usually clustered and abundant; capsules to 6 mm long; operculum conic-rostrate; calyptra cylindric, covering capsule, lacerate at base.

Habitat: Soil, rotted logs, humus, tree bases in wet forests.
Range: Eastern North America.
Gulf Coast Distribution: Florida to eastern Texas.

In its typical, robust, dendroid form, this moss is one of the easiest to recognize in our flora; sometimes, however, plants growing in flooded or very wet sites lack the dendroid habit and produce, instead, a low growth of ± simple or irregularly branched, erect secondary stems. Such forms may be difficult to recognize at first sight, but the leaf characteristics, as seen under the microscope, will permit easy identification. The name *Climacium kindbergii* (Ren. & Card.) Grout has been applied in the past to the reduced form of *C. americanum. Climacium americanum* is rather common in situations that are wet during most of the year; it sometimes forms great masses of vegetation on wet, rotted logs or peaty soil in swamps.

HEDWIGIACEAE Schimp.

Plants medium-size to rather robust, in dull, loose to dense mats; stems forked, erect-ascending; leaves crowded, appressed when dry, oblong-ovate to ovate-lanceolate, acute; cells isodiametric to elongate, papillose; costa lacking. Setae short to elongate; capsules immersed to exserted; peristome mostly lacking; operculum convex to rostrate; calyptra small and mitrate to cucullate.

This family is named in honor of Johannes Hedwig, whose book *Species Muscorum Frondosorum* (1801) is the starting point for nomenclature of mosses (except *Sphagnum*). Only one genus, *Hedwigia*, with a single species, occurs in the Gulf coastal plain.

Plants slender to robust, in small cushions or loose, spreading mats, glaucous gray-brown when dry, green when moist; stems ± rigid, irregularly branched, ascending; leaves erect to appressed or occasionally ± secund when dry, spreading when wet, oblong-ovate to ovate-lanceolate, acute to acuminate, leaf tips hyaline; margins revolute in lower part of leaf or nearly to the apex; upper cells isodiametric to slightly elongate, papillose; cells in middle of leaf base elongate and porose. Infrequently producing sporophytes in our area; sporophytes concealed in perichaetial leaves along stems; setae very short; capsules completely immersed; peristome lacking; operculum slightly convex; calyptra very small, mitrate, hairy.

HEDWIGIA P.-Beauv.
Hedwigia ciliata (Hedw.) P.-Beauv.
Figure 41, J

Habitat: Rocks, especially sandstone, tree bark, old asphalt shingle roofs, in ± open situations.
Range: Nearly worldwide.
Gulf Coast Distribution: Florida to Texas.

Hedwigia ciliata is probably more frequent along the Gulf Coast than existing collections indicate. In areas where sandstone exposures are absent, *Hedwigia* may still occur in small clumps on tree bark, walls, and old roofs. However, colonies growing in such habitats seem to be less vigorous than those on native sandstones and do not produce sporophytes. Even when sporophytes are present, they are not easy to see because they are completely hidden within the perichaetial leaves, which occur in small, spherical clusters along the sides of the stems. The absence of a costa and the hyaline leaf tips, which are easily seen with a hand lens, identify this moss at once.

CRYPHAEACEAE Schimp.

Plants slender to rather robust, ± rigid, dull to glossy, in tufts or dense or loose mats; primary stems creeping, slender, inconspicuous; secondary stems erect to pendent, simple to irregularly or subpinnately branched; leaves crowded, appressed when dry, smooth or plicate, ovate-lanceolate to oblong-lanceolate, acute to acuminate; cells slightly elongate at midleaf, smooth to papillose, quadrate in large, conspicuous areas in alar regions; costa single (occasionally short and double). Setae short; capsules mostly immersed; peristome double; operculum conic to rostrate; calyptra small, conic to cucullate, smooth to rough, sometimes hairy.

The Cryphaeaceae are a mainly tropical family of bark-inhabiting mosses. Three species in two genera occur along the Gulf Coast, mainly in humid forests. One species, *Forsstroemia trichomitria*, is sometimes very conspicuous on tree trunks and branches.

Plants dull; secondary stems slender, simple or only slightly and irregularly branched; branches remote; leaves smooth	1. CRYPHAEA
Plants glossy; secondary stems robust, subpinnately branched; branches crowded; leaves ± plicate	2. FORSSTROEMIA

1. CRYPHAEA Mohr in Web.

Plants slender, dull, dark green to yellowish, in thin colonies; secondary stems slender, short, stiffly ascending to erect, simple or with short, irregular branches; leaves tightly appressed when dry, ovate to ovate-lanceolate, acute to acuminate; cells slightly elongate, smooth to slightly papillose; costa single, ending in midleaf or above. Sporophytes usually present; setae very short; capsules immersed; endostome inconspicuous; operculum conic; calyptra very small, conic.

Leaves ovate, tips broadly acute; costa ending near midleaf	1. C. GLOMERATA
Leaves ovate-lanceolate, tips long-acuminate; costa ending in leaf apex	2. C. NERVOSA

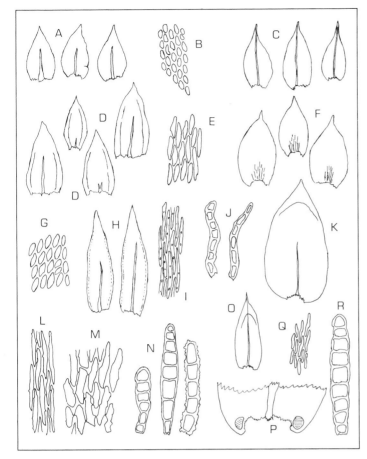

Fig. 43. *A–B*: **Cryphaea glomerata.** *A*, leaves (× 15). *B*, upper cells (× 231). *C*, **Cryphaea nervosa,** leaves (× 15). *D–E*: **Forsstroemia trichomitria.** *D*, leaves (× 15). *E*, median cells (× 231). *F–G*: **Leucodon julaceus.** *F*, leaves (× 15). *G*, median cells (× 231). *H–J*: **Leucodontopsis geniculata.** *H*, leaves (× 15). *I*, median cells (× 231). *J*, gemmae (× 231). *K–N*: **Jaegerina scariosa.** *K*, leaf (× 15). *L*, cells from upper part of leaf (× 231). *M*, cells from lower part of leaf (× 231). *N*, gemmae (× 231). *O–R*: **Pireella pohlii.** *O*, leaf (× 15). *P*, leaf base, showing regions of differentiated cells (× 55). *Q*, cells from upper part of leaf (× 231). *R*, gemma (× 231).

Colonies thin to rather dense; secondary stems slender, mostly simple, to about 1 cm tall, straight or curved; leaves ovate, apex short, acute; costa ending about midleaf. Commonly fruiting; sporophytes immersed in budlike perichaetia along the secondary stems.

> **Habitat:** Tree trunks, and on twigs and branches of trees, shrubs, and vines in moist forests; commonly encountered on fallen branches.
> **Range:** Eastern United States.
> **Gulf Coast Distribution:** Florida to Texas.

This little moss often grows quite high up on tree branches; sometimes the most luxuriant colonies found are those discovered on branches recently fallen from the tree canopy. *Cryphaea glomerata* is rather common in humid forests along the Gulf Coast. It is easily distinguished from *C. nervosa*, even in the field, by its stouter secondary stems and by its broadly acute, rather than sharply acuminate, leaves. Under the microscope the shorter costa

1. **Cryphaea glomerata** B.S.G. ex Sull.
Figure 43, A–B

is also distinctive. The simple or only slightly and irregularly branched secondary stems and nonplicate leaves distinguish it easily from *Forsstroemia trichomitria*, also common on bark. Its much smaller size and entirely immersed sporophytes help differentiate it from *Leucodon julaceus*. Also, plants of *Leucodon* and *Forsstroemia* are ± glossy, while those of *Cryphaea* are dull.

2. Cryphaea nervosa (Drumm.) C.M.
Figure 43, C

Similar to *C. glomerata* but differing most obviously in the much more slenderly pointed leaves with the costa extending well into the acumen; also the stems tend to be longer, up to 1.5–2 cm, and more slender, and are more often branched than in *C. glomerata*. Commonly fruiting; sporophytes as in *C. glomerata*.

> **Habitat:** Similar to that of *C. glomerata* but generally in moister sites.
> **Range:** Southeastern United States.
> **Gulf Coast Distribution:** Florida to eastern Texas.

Easily told from *C. glomerata* by the characteristics cited in the key and above. See comments under *C. glomerata*. This species is less common than *C. glomerata*; however, it is often found in abundance in favorable sites. It is not uncommon to find both species growing together in the same habitat.

2. FORSSTROEMIA Lindb.

Forsstroemia trichomitria
(Hedw.) Lindb.
Figure 43, D–E

Plants slender to ± robust, glossy, in tufts or dense mats; secondary stems erect to pendent, usually highly subpinnately branched, branches mostly crowded; leaves erect to appressed when dry, ± plicate, oblong-ovate to ovate-lanceolate, acute, margins somewhat reflexed; cells smooth, fusiform; costa slender, single, reaching above midleaf or shorter, often short and double or nearly absent. Sporophytes usually produced in abundance; setae very short; capsules immersed to slightly exserted; peristome fragile, endostome indistinct; operculum obliquely rostrate; calyptra cucullate, hairy.

> **Habitat:** Tree bark, rarely rock, in mesic forests and along streams.
> **Range:** Eastern North America, south to northeastern Mexico.
> **Gulf Coast Distribution:** Florida to Texas.

This moss is common and abundant along the Gulf Coast wherever there are forests with adequate humidity. It forms great, shaggy colonies on trees, especially ironwood and beech but also many others, in favorable situations along streams and in river bottoms. *Forsstroemia trichomitria* differs from *Leucodon* and *Cryphaea*, among other ways, in having the secondary stems rather regularly and densely branched. The bark-dwelling moss in our area which *Forsstroemia* most closely resembles in the field is *Pireella pohlii*; see comments under that species for distinctions.

LEUCODONTACEAE Schimp.

Plants medium-size to robust, green to yellowish green or brownish; primary stems creeping; secondary stems ascending to pendent, simple or branched; leaves crowded, ovate to ovate-lanceolate, smooth or plicate; cells oval to somewhat elongate, smooth or papillose, quadrate in conspicuous areas at basal angles; costa single, short and double, or lacking. Setae short or elongate; capsules immersed to exserted; peristome mostly double, endostome sometimes reduced or lacking; operculum conic to rostrate; calyptra cucullate, naked or somewhat hairy.

The Leucodontaceae are primarily bark-dwelling mosses. Two genera have been recorded from our area. One species is very common essentially throughout the Gulf Coast.

Leaf cells smooth, or somewhat bulging dorsally near the leaf tip	1. LEUCODON
Leaf cells distinctly papillose	2. LEUCODONTOPSIS

Plants ± glossy, light to dark green or brownish, in thin, spreading mats or dense, shaggy colonies; primary stems slender; secondary stems stout, erect-ascending to pendent, simple or with occasional branches; leaves crowded, appressed when dry, broadly ovate, abruptly acute to acuminate, not plicate; cells slightly elongate, ± protruding dorsally in upper part of leaf; costa lacking. Sporophytes commonly produced; setae short to elongate; capsules emergent to long-exserted, ovoid, constricted at mouth; operculum obliquely rostrate; calyptra cucullate, naked; peristome double, but the endostome represented by a low, pale membrane.

1. LEUCODON Schwaegr.

Leucodon julaceus (Hedw.) Sull.
Figure 43, F–G

> **Habitat:** Trees, logs, stumps, rarely on rock; mesic to rather dry, open forests.
> **Range:** Eastern North America; northern Mexico.
> **Gulf Coast Distribution:** Florida to Texas.

Leucodon julaceus is one of our commonest bark-dwelling mosses. The plants are quite variable in aspect, depending on the substrate and the age and condition of the colony. Young colonies, or those growing on poor substrates, are thin, with remote, short, secondary stems arising from the creeping primary stem. Mature colonies in favorable sites may be very dense and shaggy, with the stout, elongate secondary stems pendent and curved at the tips. Abundant sporophytes are produced on most colonies in favorable situations. This moss often grows high up in trees, but the best growth seems to be lower down, on the main trunk and larger branches. The highest colonies are mostly small, scraggly, and sterile. *Cryphaea glomerata* and *C. nervosa* have a similar growth habit, but their secondary stems are much thinner, the plants are dull, and the sporophytes are completely immersed. Under the microscope the absence of a costa will distinguish *Leucodon* from *Cryphaea*.

LEUCODON BRACHYPUS Brid. was reported from our area, in Mobile County, Alabama, by Wilkes (1965). It differs most obviously from *L. julaceus* in having decidedly plicate leaves which are erect-spreading when moist, rather than wide-spreading as in *L. julaceus*. Although this species is widespread in North America, it is doubtful that it occurs so far south. I have not been able to see the specimen upon which the Alabama report was based.

2. LEUCODONTOPSIS Ren. & Card.

Leucodontopsis geniculata
(Mitt.) Crum & Steere
Figure 43, H–J

Plants medium-size, yellowish green to brownish, in thin to dense colonies; primary stems slender, covered with brown rhizoids; secondary stems erect-ascending, simple or little-branched; leaves loosely appressed, plicate, narrowly lanceolate to ovate-lanceolate, acuminate, margins recurved; cells fusiform to linear, papillose, quadrate or ± transversely elongate in conspicuous areas at basal angles; costa extending to midleaf or beyond, or lacking; short, uniseriate gemmae usually present in leaf axils. Not producing sporophytes in our area.

> **Habitat:** Tree trunks and branches; fence posts; ± open, humid forests.
> **Range:** Tropical America; Africa.
> **Gulf Coast Distribution:** Florida.

This species has been reported in our area from Jackson County, Florida. It is fairly common in southern Florida and could turn up elsewhere in our area, especially in the eastern part. The distinctly papillose leaf cells distinguish *L. geniculata* from similar corticolous mosses such as *Cryphaea* and *Leucodon*. I have not seen the specimen upon which the western Florida report was based, and it is possible that the report is based on a misidentification.

PTEROBRYACEAE Kindb.

Plants medium-size to robust, green to yellowish green; primary stems creeping, rhizomelike; secondary stems erect to horizontal or pendent, simple or branched, tumid; leaves concave, loosely imbricate, sometimes in distinct rows, mostly ovate-lanceolate, pointed, often with uniseriate gemmae in the axils; cells rhombic to linear, smooth to papillose, often somewhat differentiated in alar regions; costa single or short and double, sometimes nearly absent. Sporophytes not produced in our area.

The Pterobryaceae are mainly tropical, corticolous mosses, often common and abundant in the warm parts of the world. Representatives of two genera reach our area. Although similar in habitat and general aspect to members of the Cryphaeaceae and Leucodontaceae, mosses of this family can usually be recognized in the field by their very concave leaves, which cause the stems to be soft and tumid.

Secondary stems sparse, simple, remote on primary stems	1. JAEGERINA
Secondary stems abundant, freely branched, crowded on primary stems	2. PIREELLA

Plants medium-size, green to yellowish green; primary stems naked or with scalelike leaves, inconspicuous; secondary stems erect to pendent, to about 3 cm long but mostly shorter, densely foliated, glossy, simple, with stipelike base bearing reduced leaves; leaves loosely spreading, concave, ovate-lanceolate, short-pointed, apex often rather blunt; cells fusiform to short-linear, smooth, a few quadrate to short-rectangular and porose at basal angles; short, uniseriate, smooth or papillose gemmae usually present in leaf axils; costa single, short and double, or virtually absent.

1. JAEGERINA C.M.

Jaegerina scariosa (Lor.) Arzeni
Figure 43, K–N

> **Habitat:** Tree bark in dense, humid forests.
> **Range:** Southeastern United States; West Indies; Central America; Mexico; northern South America.
> **Gulf Coast Distribution:** Florida; southeastern Louisiana.

This is a very distinctive species which cannot easily be confused with any other in our area. *Leucodon julaceus* has somewhat the same habit, but its leaves are appressed when dry, rather than loosely spreading, and have conspicuous areas of quadrate cells at the basal angles. *Cryphaea* is much more slender than *Jaegerina*. This species is quite rare along the Gulf Coast but is probably more widely distributed than existing specimens indicate. It is not easy to find because it commonly grows with other bryophytes on trees in densely vegetated habitats that are difficult to penetrate. Reproduction and dispersal occur in our area by the usually abundant gemmae, which may be short and smooth or longer and papillose, or smooth in the lower portion and papillose above.

Plants medium-size, green to yellowish green; primary stems with scalelike leaves, creeping and inconspicuous; secondary stems erect to pendent, to about 4 cm long, with stipelike base bearing reduced leaves, densely ± pinnately branched, both stems and branches densely foliated and tumid, glossy; leaves deeply concave, about 1.5 mm long, acuminate, ovate-lanceolate; cells

PIREELLA Card.

Pireella pohlii (Schwaegr.)
Card.
Figure 43, O–R

smooth, fusiform to short-linear, a few at basal angles quadrate and colored; basal angles of leaves somewhat auriculate and clasping the stem; costa single; short uniseriate gemmae usually abundant in leaf axils; secondary stems and branches sometimes becoming long, slender, and flagelliform.

Habitat: Tree bark in humid hardwood forests.
Range: Southeastern United States; tropical America.
Gulf Coast Distribution: Florida; Mississippi; Louisiana.

Pireella pohlii has not yet been reported from Alabama but surely occurs there; it is probably of general occurrence along the Gulf Coast from Florida to south central Louisiana and could well turn up in southeastern Texas. *Forsstroemia trichomitria* is similar in habit and habitat but lacks gemmae, is more slender, and has tightly appressed leaves when dry; it is also commonly found fruiting while *Pireella* is not known to produce sporophytes in our area. The leaves of *Forsstroemia* often appear to be plicate when dry and are not nearly so concave and glossy as those of *Pireella*. This handsome moss forms large, shaggy colonies on hardwood trees in deep forests and along river banks. It is fairly common in favorable sites in the eastern part of our area.

PIREELLA CYMBIFOLIA (Sull.) Card. was reported from Louisiana long ago but apparently does not actually occur there. I have not seen an authentic specimen. It is known in the United States only from southern Florida. It is similar in most respects to *P. pohlii* but can be easily distinguished by its numerous quadrate alar cells which form conspicuous areas and by its nonauriculate basal leaf angles. It is also less regularly branched than *P. pohlii*.

METEORIACEAE Kindb.

Plants slender to moderately robust, green to yellowish green or brownish; primary stems slender, creeping; secondary stems mostly elongate, pendent, freely branched, often forming tangled masses; leaves mostly ovate-lanceolate, acuminate, often concave, frequently filiform at the tips; cells elongate, often papillose; costa single or lacking. Setae short; capsules exserted, erect, symmetric; peristome double; operculum short-rostrate; calyptra often hairy.

The Meteoriaceae are mainly plants of the tropics, with a few representatives that extend into adjacent regions. Virtually all the members of this large family are epiphytes, growing in intricate masses dangling from twigs and branches. Two species, representing different genera, occur along the Gulf Coast.

Plants typically creeping in thin, straggly mats on bark, moderately
robust; secondary stems and branches densely foliated; leaf cells
arranged in rows diverging from costa toward margins .... 1. PAPILLARIA

Plants typically pendent from twigs, delicate and slender; secondary stems and branches mostly loosely foliated; leaf cells not in
rows diverging from costa .... 2. BARBELLA

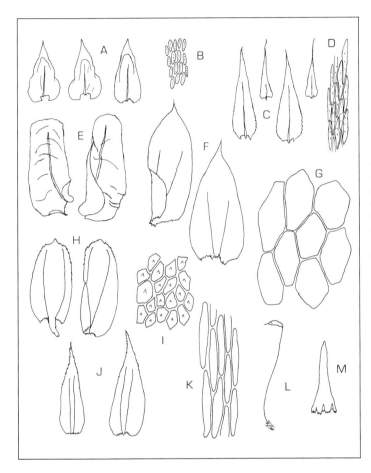

Fig. 44. *A–B*: **Papillaria nigrescens.** *A,* leaves (× 15). *B,* upper cells (× 231). *C–D*: **Barbella pendula.** *C,* leaves (× 15). *D,* upper cells (× 231). *E,* **Neckeropsis disticha,** leaves (× 15). *F–G*: **Cyclodictyon varians.** *F,* leaves (× 18). *G,* upper cells (× 231). *H–I*: **Schizomitrium pallidum.** *H,* leaves (× 27). *I,* upper cells (× 395). *J–M*: **Hookeriopsis heteroica.** *J,* leaves (× 25). *K,* upper cells (× 371). *L,* sporophyte (× 3). *M,* calyptra (× 15).

Plants dull, green to brownish green, slender to moderately robust;
primary stems subpinnately branched; branches terete, short or
somewhat elongate, ± erect; plants often with flagelliform, naked
or microphyllous, branches; leaves usually crowded, ± appressed
when dry, deeply concave, ovate-lanceolate to triangular, somewhat auriculate, short-pointed to long, filiform-acuminate, margins
subentire to serrulate; median cells elongate, pointed at ends, bearing several papillae in a row on each surface, cells arranged in rows
diverging from the costa towards the margins; costa ending at midleaf or in base of acumen. Not fruiting in our area.

1. PAPILLARIA (C.M.) C.M.

Papillaria nigrescens (Hedw.)
Jaeg. & Sauerb.
Figure 44, A–B

Habitat: Bark of trees and vines in low forests.

Range: Southeastern United States; tropical America; Galapagos.

Gulf Coast Distribution: Florida; Mississippi; Louisiana; always near the coast. Not yet reported from the Florida Panhandle or from Alabama but probably present there.

This moss shows its relationship to *Barbella pendula* by the similarity in leaf shape and papillose cells, but the two species are not at all similar in most respects and are not usually mistaken for one another. *Papillaria nigrescens* is never abundant or common in our area, but I once found a considerable quantity of it in a rodent nest built in the leaf bases of a palmetto plant. I was unable to find the *Papillaria* otherwise in the forest in the general vicinity of the nest, even though I made a careful search for it. Apparently the animal had selectively harvested the *Papillaria* for use in its nest because, although there were other corticolous mosses present in abundance, the evidently scarce *Papillaria* was the only moss used in the nest. See the discussion under *Barbella pendula* for further comments on *Papillaria nigrescens*.

2. BARBELLA Fleisch. in Broth.

Barbella pendula (Sull.) Fleisch.
Figure 44, C–D

Rather delicate, slender, bright green plants, usually pendent in tangled wisps; pendent stems greatly elongated, bearing short, irregularly placed branches; leaves lanceolate, slenderly long-acuminate, some with a long, delicate, filiform acumination of elongated cells joined end-to-end; leaf margins subentire to serrulate; median leaf cells elongate, bearing a median row of several small papillae dorsally, alar cells quadrate in distinct groups; some leaves very narrow and appressed, others broader and spreading; costa ending in base of acumen. Capsules ovoid, dark; operculum obliquely short-rostrate; peristome pale; calyptra small, naked, split on one side in basal portion. Rarely fruiting.

Habitat: Creeping or pendent on twigs and stems of saplings and cane in deep, shaded ravines in deciduous and mixed forests.

Range: Southern United States (Louisiana and Mississippi); Mexico; Japan; southeastern China.

Gulf Coast Distribution: West Feliciana Parish, Louisiana, and Wilkinson County, Mississippi.

Barbella pendula is a beautiful plant, and the only moss with a pendent habit in our area. It was once more abundant and widely distributed in Louisiana (and possibly Mississippi) than at present. Presumably, agricultural drainage and clearing, deforestation, and modification of stream channels for flood control have reduced the habitat available for *Barbella*. Collections from the last century indicate that this moss once grew in great abundance in some sites where it is apparently absent today. The report of *B. pendula* from

Florida, discussed by Breen (1963), was based on specimens of *Papillaria nigrescens*. *Barbella pendula* is known in the United States today only from a few deep, wooded ravines in Louisiana and from an adjacent area in Mississippi.

The range of *Barbella pendula* indicates that it has been a resident of North America for a very long time, probably originally as a member of an ancient, warm-climate, circumboreal, Tertiary flora whose components were dispersed southward in some areas, or extirpated in others, by a cooling climate. Vascular plants such as *Magnolia grandiflora* and sassafras were also members of this ancient flora, as were the mosses *Syrrhopodon texanus* and *S. prolifer*, for example. In contrast, the other Gulf Coast member of the moss family Meteoriaceae, *Papillaria nigrescens*, is more likely to be a relatively recent arrival here, having gained footholds along the Gulf Coast in Florida, Mississippi, and Louisiana through recent migration(s) from the tropics, where it is widespread and abundant.

NECKERACEAE Schimp.

Plants mostly robust, often glossy; primary stem creeping; secondary stems spreading to pendent, irregularly branched, strongly flattened; leaves complanate, mostly broad and short-pointed, often transversely undulate; cells smooth, short above, linear below; costa single or short and double. Sporophytes terminal on short branches of secondary stems; setae short; capsules immersed to exserted; peristome double; calyptra cucullate.

The Neckeraceae are a fairly large family, with best representation in the warmer parts of the world. Only a single species, in the genus *Neckeropsis*, has been reported from our area.

NECKEROPSIS Reich.
Neckeropsis undulata (Hedw.) Reichardt
Figure 44, E

Plants green to yellowish green, in thin, loose colonies, moderately robust; primary stems slender, creeping, secondary stems strongly flattened, spreading or pendent, irregularly branched; leaves in four rows, two each on opposite sides of the stem, but appearing as though in two rows, broadly rectangular from an asymmetric base, truncate at apex, sometimes bluntly apiculate, usually strongly transversely undulate; median and upper cells firm, thick-walled, irregularly rhomboidal, lower cells becoming linear; margins plane, entire, or irregularly serrate at apex; costa slender, ending well below leaf apex. Sporophytes immersed or nearly exserted, closely enveloped by the bristlelike inner perichaetial leaves, terminal on very short lateral branches; capsules oblong-cylindric; peristome teeth pale, slenderly triangular, segments yellow, slender, anastomosing; spores smooth; operculum obliquely slenderly rostrate; calyptra naked or with a few hairs.

Habitat: Mostly corticolous, sometimes on stumps, logs, or rock; in humid forests.

Range: Southeastern United States; tropical and subtropical America.

Gulf Coast Distribution: Texas (?)

This distinctive moss was collected long ago in Texas by Charles Wright (Whitehouse and McAllister 1954) but the locality is not known today. It has not been found in our area since. It is fairly frequent outside of our area in Florida, especially south of the Suwannee River. The plants are rather similar in aspect and habit to some leafy liverworts, especially *Plagiochila*, but the undulate leaves, for example, are characteristic. It cannot be confused with any other of our mosses. *Neckeropsis undulata* could well turn up anywhere along the Gulf Coast, especially in the eastern part of our range. I have not seen the specimen upon which this report from Texas is based, and it is possible that the report is in error.

HOOKERIACEAE Schimp.

Plants small to robust, rather soft, shriveled when dry; stems mostly freely branched, ± prostrate (in our species); leaves often ± complanate, somewhat dimorphic, the lateral ones differing from the dorsal ones, ovate-lanceolate, often acuminate, frequently bordered with elongate cells; cells smooth to papillose, often lax, isodiametric or elongate, not much differentiated in alar regions; costa double (in our species). Setae elongate, smooth or papillose; capsules erect to inclined or pendulous; peristome double; operculum rostrate; calyptra naked or variously hairy or scaly, lobed at base, sometimes deeply split on one side, conic to mitrate-campanulate.

The Hookeriaceae are primarily plants of the tropics, where they are diverse and abundant. Four species, representing different genera, have reached our area, where they occur mainly in the southern portion of our range.

1. Leaf cells unipapillose; leaf apex rounded to apiculate; seta papillose — 2. SCHIZOMITRIUM
1. Leaf cells smooth; leaf apex acute; seta smooth or papillose — 2
  2. Plants robust; primary stems creeping, inconspicuous; secondary stems erect or ascending, not or little branched; leaf margins coarsely toothed above; seta papillose; calyptra hairy — 4. LEPIDOPILUM
  2. Plants slender; stems all prostrate, freely branched; leaf margins entire or serrate, not coarsely toothed; seta smooth; calyptra naked — 3

3. Median cells large, essentially isodiametric; leaf apex with short, entire acumination
3. Median cells narrow, much longer than wide; leaf apex with long, serrate acumination

Plants small, in pale green, intricate mats, with metallic luster when dry; stems much branched; leaves complanate, shriveled when dry, ovate to broadly ovate, with short acumination, marginal row of cells very narrow, forming distinct border, margins entire or slightly toothed towards apex; median cells large, lax, approximately hexagonal; costae divergent, ending in base of acumen. Setae smooth; capsules inclined, ± asymmetric; operculum slenderly rostrate; calyptra naked, lobed at base.

> **Habitat:** In wet places or aquatic; on rocks, soil, and humus; springs, stream banks, and ravines in forests.
> **Range:** Southeastern United States; West Indies; Central America; northern South America.
> **Gulf Coast Distribution:** Louisiana.

Cyclodictyon varians is beautiful to see under the microscope. The large, lax cells and the two costae are striking. This species is known from a number of collections outside of our area in Florida. It is known in Louisiana from two collections in Iberia Parish in the south central part of the state, where it was found in a deep ravine on Avery Island (an emergent salt dome). It might well be expected to turn up near the coast in Mississippi and Alabama, as well as in the Panhandle counties of Florida.

1. CYCLODICTYON Mitt.

Cyclodictyon varians (Sull.) Kuntze
Figure 44, F–G

Plants small, in thin, dull green mats; stems freely branched; leaves ± complanate, shriveled when dry, ovate-oblong, apex bluntly rounded, often somewhat apiculate, margins serrulate, especially above; median cells isodiametric or a little longer than wide, each with a single sharp papilla on both surfaces; costae parallel or slightly convergent above, ending in leaf apex, toothed dorsally above. Setae papillose; capsules ovoid, inclined to pendulous; operculum abruptly short-rostrate; calyptra naked, deeply lobed at base, mitrate-campanulate, sometimes split on one side from base.

> **Habitat:** Moist sites, sometimes ± aquatic; rock, rotted wood, soil; springs, stream banks, ravines, in forests.
> **Range:** Southeastern United States; Mexico; Central America; West Indies; South America.
> **Gulf Coast Distribution:** Louisiana.

This interesting little moss is better known under its former name, *Callicostella pallida*. It has not yet been found in the Panhandle counties of Florida or in Alabama or Mississippi, but should

2. SCHIZOMITRIUM B.S.G.

Schizomitrium pallidum (Hornsch.) Crum & Anderson
Figure 44, H–I

Callicostella pallida (Hornsch.) Ångstr.

eventually turn up there. Like the other Gulf Coast members of the family Hookeriaceae, *S. pallidum* is probably a relatively recent arrival from the tropics.

3. HOOKERIOPSIS (Besch.) Jaeg. & Sauerb.

Hookeriopsis heteroica Card.
Figure 44, J–M

Plants small to medium-size, glossy, light to brownish or golden green, in soft, dense mats; stems freely branched; leaves somewhat complanate, long and slenderly pointed, ovate-lanceolate, acumen serrate on margins; cells narrow, elongate, pointed at ends, smooth; costae parallel, toothed dorsally above, ending in base of acumen. Setae smooth; capsules dark, ovoid, inclined; operculum obliquely short-rostrate; calyptra naked, lobed at base, often split on one side.

> **Habitat:** Wet places; on rocks, soil, tree bases, rotted wood; springs, banks; in forests.
> **Range:** Southeastern United States; Mexico; Cocos Islands.
> **Gulf Coast Distribution:** Washington Parish, Louisiana; Wilkinson County, Mississippi.

This handsome moss was unknown in the United States until 1969, when I discovered it at the Mississippi site. Since then I have found it at two locations in Louisiana, not far from the Mississippi station but in a different drainage basin. It is likely that this species will turn up eventually elsewhere along the Gulf Coast, especially toward the east. In the field, *Hookeriopsis* vaguely resembles an *Amblystegium*; under the microscope the two costae identify it immediately as a member of the Hookeriaceae.

4. LEPIDOPILUM (Brid.) Brid.

Lepidopilum polytrichoides (Hedw.) Brid.
Figure 45, A–C

Plants robust, somewhat glossy; primary stems creeping on bark, inconspicuous; secondary stems large, erect or ascending, not or little branched, elongate; leaves complanate, dorsal leaves somewhat differentiated from lateral, obovate to lanceolate, apiculate, apiculation often twisted, margins coarsely toothed by sharp, projecting cells; median cells several times longer than wide, pointed on ends, rather lax, smooth; costae divergent, ending about midleaf. Setae papillose; capsules erect; operculum conic-rostrate; calyptra campanulate with sparse hairs and scales, lobed at base.

> **Habitat:** On trees in forests.
> **Range:** Southeastern United States; tropical America.
> **Gulf Coast Distribution:** Escambia County, Florida.

This moss is included on the basis of a specimen collected long ago at Pensacola, Florida. The species has not been found since on the Gulf Coast. It is possible that the specimen was collected elsewhere and mislabeled as to locality; on the other hand, *L. polytrichoides* is very widespread in tropical America and could well have reached the Gulf Coast, as did other members of the Hookeriaceae, but failed to persist or has been overlooked.

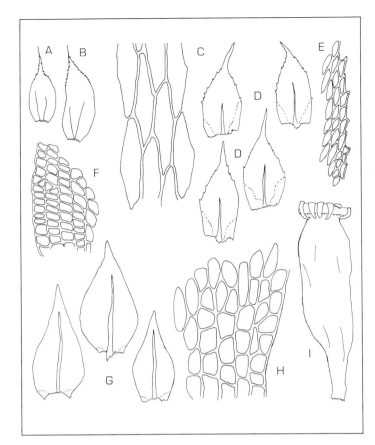

Fig. 45. *A–C:* **Lepidopilum poly-trichoides.** *A,* dorsal leaf (× 5). *B,* lateral leaf (× 5). *C,* upper cells (× 231). *D–F:* **Fabronia ciliaris.** *D,* leaves (× 55). *E,* upper margin and adjacent cells (× 231). *F,* cells in alar region (× 231). *G–I:* **Ana-camptodon splachnoides.** *G,* leaves (× 27). *H,* cells in alar region (× 231). *I,* capsule, showing re-curved peristome teeth (× 34).

FABRONIACEAE Schimp.

Small, soft, slender, often glossy plants, in thin, intricate mats; dark to pale green; stems prostrate, freely branched; branches ascending or prostrate, usually terete; leaves soft, somewhat concave, median cells slightly elongate, pointed at ends, mostly smooth (with small but distinct papillae in *Schwetschkeopsis*), basal and alar cells often quadrate, sometimes forming conspicuous areas; costa lacking, or single and ending at or beyond midleaf. Setae short to elongate; capsules exserted, erect and symmetric; peristome single or double; operculum conic to short-rostrate, rostrum often oblique; calyptra cucullate, naked.

The Fabroniaceae are a small family of rather delicate, in-conspicuous plants mostly growing on bark. One of them, *Clasmatodon parvulus*, is exceedingly common and abundant throughout our area, but is rarely conspicuous due to its small size.

1. Leaves ecostate, margins serrulate; median cells papillose
   dorsally .... 3. SCHWETSCHKEOPSIS
1. Leaves costate, margins entire or toothed; median cells smooth
   dorsally 2
    2. Leaf margins mostly toothed above .... 1. FABRONIA
    2. Leaf margins entire or slightly serrulate 3
3. Plants very slender; stems and branches tightly foliated when
   dry, the leaves appressed; peristome teeth erect or incurved
   when dry, quickly deciduous; capsule not constricted under
   the mouth when dry; flagelliform branches often present ... 4. CLASMATODON
3. Plants more robust; stems and branches loosely foliated when
   dry, the leaves spreading-homomallous; peristome teeth
   strongly recurved when dry, persistent; capsule strongly constricted below the mouth when dry; flagelliform branches
   lacking .... 2. ANACAMPTODON

1. FABRONIA Raddi

Slender, soft, delicate plants in intricate, pale green, thin mats; stems mostly irregularly branched; stems and branches usually fuzzy or hairy in appearance, especially at tips, due to the delicate, spreading leaf tips; leaves broadly ovate to lanceolate, with long, slender acumen of narrow, very elongate cells, margins usually prominently dentate or ciliate-dentate with large, delicate, project-ing cells, the teeth sometimes extending nearly to leaf base, or scarce or virtually absent; median cells lax, longer than wide, pointed at ends, smooth, alar cells quadrate, numerous; costa end-ing about midleaf. Setae short; capsules pyriform to ovoid, short; peristome single; operculum apiculate to very short-rostrate.

Fabronia is a small genus of bark- and rock-dwelling plants. Al-though none of the species has yet been reported from the area covered by this manual, three species occur around the edges of the manual range, and one or more of them could be found in the Gulf Coast area covered here. The species are not particularly dis-tinct from one another.

1. Leaves entire or nearly so; southeastern .... F. RAVENELII Sull.
1. Leaves dentate to ciliate-dentate; southwestern or widespread ... 2
    2. Leaves lanceolate to narrowly lanceolate, ± evenly dentate;
     southwestern .... F. WRIGHTII Sull.
    2. Leaves broadly ovate to ovate-lanceolate; often irregularly
     toothed; widespread in eastern North America .... F. CILIARIS (Brid.) Brid.

Fabronia ciliaris is illustrated in Figure 45, D–F.

The distinctions described for these three species are not out-standing. However, plants of the genus Fabronia are easy to recognize by the dentate margins and the long, slender acumination of the leaves. The plants are commonly found with sporophytes and the short, erect, ± pyriform capsules with apiculate or very short-rostrate operculum are also helpful in recognition. In *Clasmatodon*, the leaves are never long and slenderly acuminate; the acumination is not formed of narrow, elongate cells; and the margins, although sometimes serrulate, are never dentate with large, delicate, projecting cells. In the field, *Clasmatodon* stems and branches are not fuzzy in appearance because the leaves are appressed and lack the long spreading acumination of *Fabronia*. The leaf tips of *Clasmatodon* may be slightly spreading, but they are not hairlike as in *Fabronia*. *Clasmatodon* and *Fabronia* occupy essentially the same types of habitats.

Plants small, in usually loose, dark green mats; stems freely branched; branches ascending, short; leaves loosely spreading-ascending when dry, often homomallous, ovate-lanceolate, acuminate, margins entire; median cells several times longer than wide, ends pointed, smooth, a few cells quadrate in alar regions, marginal row of cells somewhat differentiated; costa ending in base of acumen. Sporophytes stout; setae elongate, capsules strongly contracted below the mouth when dry; peristome double, teeth recurved, bending out and down over rim of mouth when dry, persistent, segments of endostome small, slender, erect or ± incurved when dry; operculum conic to shortly and obliquely rostrate.

2. ANACAMPTODON Brid.

Anacamptodon splachnoides
(Froel. ex Brid.) Brid.
Figure 45, G–I

Habitat: Wet, damp, or seeping places on trees, stumps, rarely on logs; knotholes, cracks, crotches on broadleaf trees, rarely on bald cypress; forests.

Range: Eastern North America; Central Europe.

Gulf Coast Distribution: Florida to eastern Texas.

This interesting moss has the common name "knothole moss" due to its propensity for inhabiting wet holes in trees; about half of the collections I have seen were taken from knotholes, but the moss also grows freely on the wet tops of stumps and around seeping fissures in trees, so it is not by any means restricted to knotholes. Nonetheless, if one is searching for *Anacamptodon*, knotholes are a prime place to look.

When it is found with sporophytes, *Anacamptodon* is easy to identify due to the robust sporophytes with erect capsules, strong contraction below the mouth, and the strikingly recurved peristome teeth. No other moss in our area has the above features and grows in similar habitats. When sterile, *Anacamptodon* can still be easily identified by its habitat, the rich, dark green color, and the leaves. Although it may vaguely resemble an *Amblystegium*, *Anacamptodon* is a larger plant and occupies a different habitat.

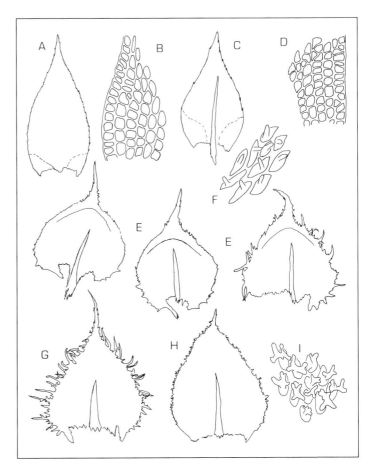

Fig. 46. *A–B*: **Schwetschkeopsis fabronia.** *A*, leaf (× 55). *B*, cells in alar region (× 231). *C–D*: **Clasmatodon parvulus.** *C*, leaf (× 55). *D*, cells in alar region (× 231). *E–F*: **Thelia hirtella.** *E*, leaves (× 55). *F*, median cells (× 385). *G*, **Thelia asprella,** leaf (× 55). *H–I*: **Thelia lescurii.** *H*, leaf (× 55). *I*, median cells (× 385).

3. SCHWETSCHKEOPSIS Broth.

Schwetschkeopsis fabronia
(Schwaegr.) Broth.
Figure 46, A–B

Schwetschkeopsis denticulata (Sull.) Broth.

Very similar in most respects to *Clasmatodon parvulus*; slender, pale green, in thin or dense mats; stems prostrate, freely subpinnately branched, often ± flattened, usually with abundant flagelliform branches; stems and branches usually all pointing in the same direction; branches prostrate or ascending; leaves concave, often ± complanate, ovate-lanceolate or broader, shortly acuminate, apex often oblique, margins serrate all around, median cells longer than wide, pointed at ends, finely but distinctly papillose dorsally at ends, especially in upper part of leaf; costa lacking. Not fruiting in our area.

> **Habitat:** Almost exclusively on bark, especially on tree bases; occasionally on rock; forests.
> **Range:** Eastern United States; Asia.
> **Gulf Coast Distribution:** Florida to eastern Texas.

Schwetschkeopsis fabronia is not nearly as common or abundant as *Clasmatodon parvulus*, with which it often grows. It can usually be told from *Clasmatodon* in the field because plants of *Schwetschkeopsis* are generally lighter in color and more consistently "combed" in appearance. Furthermore, whereas the stems and branches of *Clasmatodon* are mostly terete and all about the same diameter, those of *Schwetschkeopsis* are often flattened, with the stems conspicuously wider than the branches. Under the microscope the absence of a costa and the serrate leaf margins and papillose cells distinguish it at once from *Clasmatodon*. I have never seen *Schwetschkeopsis* from the Gulf Coast with sporophytes, although it apparently does fruit in the northern part of its range in the United States.

Plants very slender, light to dark green, in usually thin, often straggly mats; stems prostrate, freely branched; slender, elongated, flagelliform branches often present; branches short, ascending to prostrate; stems and branches often all pointing in the same direction; leaves usually appressed with only the tips ± spreading sometimes, ovate-acuminate to broadly ovate with short, acute apex, or lanceolate, apex often oblique; margins entire or slightly serrulate above; median cells longer than wide, pointed at ends, marginal row of cells slightly differentiated, alar cells quadrate, numerous, forming conspicuous areas; costa ending about midleaf. Sporophytes small; setae short; capsules constricted at mouth, ovoid; peristome single, pale, delicate and soon falling, teeth pointing inwards; operculum sharply and obliquely short-rostrate.

> **Habitat:** Bark; trees, logs, stumps; also on rocks; forests.
> **Range:** Southeastern United States; Europe.
> **Gulf Coast Distribution:** Florida to Texas.

Clasmatodon parvulus is one of our commonest mosses and is essentially ubiquitous throughout the Gulf Coast. Although it is similar in appearance in the field to *Fabronia*, the toothed leaf margins of the latter will separate the two easily under the microscope. *Fabronia* does not occur along the coast over most of our area. *Clasmatodon* often grows with the likewise common *Sematophyllum adnatum*, which shares the same habitat and also has erect capsules. In *Sematophyllum*, however, the leaves are looser and mostly homomallous, and the peristome is double and conspicuous, even as seen with the hand lens. Under the microscope the differences in the leaves of the two species are apparent.

4. CLASMATODON Hook. & Wils. ex Wils.

Clasmatodon parvulus
(Hampe) Hook. & Wils. ex Sull.
Figure 46, C–D

LESKEACEAE Schimp.

Plants slender to rather robust, mostly dull, firm or wirey, dark green to yellowish or brownish, sometimes appearing glaucous, in thin to dense, often springy mats; paraphyllia present or lacking, not usually abundant or conspicuous; stems prostrate, creeping or arched-ascending, mostly irregularly branched (regularly pinnate in some *Thelia*); branches spreading or ascending; stem and branch leaves mostly not differentiated, mostly concave, apex blunt to acute or acuminate; leaf cells mostly uni- or pluripapillose (smooth in *Herpetineuron*), mostly ± isodiametric; costa single, strong, ending at midleaf or extending into acumen; perichaetial leaves differentiated. Setae elongate; capsules mostly erect and symmetric, slightly curved in some; peristome double (endostome sometimes reduced); operculum conic to rostrate; calyptra cucullate, naked.

A large and diverse family, sometimes united with the next, the Thuidiaceae. Viewed as a whole, this family is best recognized in the field by the dull, usually freely branched plants of often dark color. The erect, symmetric capsules and usually scanty (if present) paraphyllia help distinguish the Leskeaceae from the Thuidiaceae. Species representing six genera occur in our area.

1. Leaf cells smooth; leaf apex coarsely toothed	6. HERPETINEURON
1. Leaf cells uni- or pluripapillose; leaf apex entire or toothed	2
2. Leaf margins ciliate or spinose-papillose-toothed; leaf cells each with large, simple or branched dorsal papilla	1. THELIA
2. Leaf margins entire to ± serrulate; leaf cells with low papillae	3
3. Leaves abruptly acuminate into a pale, awnlike point; branches bearing clusters of small, leafy propagula in axils of upper leaves; cells distinctly unipapillose	2. LINDBERGIA
3. Leaves gradually narrowed to acute or blunt tips; leafy propagula lacking; cells uni- to pluripapillose, or merely bulging	4
4. Plants slender; marginal cells of leaves conspicuously bulging-papillose; tips of older leaves usually broken off	4. HAPLOHYMENIUM
4. Plants mostly coarser; marginal cells of leaves not bulging-papillose; leaves mostly intact	5
5. Leaf cells bulging or obscurely unipapillose; stem leaves not much different from branch leaves	3. LESKEA
5. Leaf cells pluripapillose; stem leaves much smaller than branch leaves	5. ANOMODON

1. THELIA Sull.

Plants small to moderate size, in thin or dense mats or clumps, green to greenish gray or glaucous; stems pinnately or irregularly branched, creeping or ascending, bearing paraphyllia; branches mostly short, ascending, terete to julaceus, simple or forked; leaves closely imbricate, concave, broadly triangular-ovate, rather abruptly apiculate, apiculation appressed or spreading; margins variously toothed and ciliate; median cells two to three times longer than wide, pointed at ends, stoutly unipapillose, the papilla simple or branched; costa ending at or above midleaf. Capsules erect, cylindric; peristome pale; operculum conic-rostrate.

Thelia is a small genus of three species, all of which are restricted to North America. All three species occur along the Gulf

Coast and are usually easy to recognize in the field by their habitat and habit.

1. Dorsal papillae simple; plants mainly on bark of trees and stumps — 1. T. HIRTELLA
1. Dorsal papillae branched; plants mainly on tree bases, soil, and rock — 2
   2. Stems pinnately branched, prostrate; branches slender, simple, short; leaf apex slenderly long-acuminate; plants in tight mats, mainly on tree bases — 2. T. ASPRELLA
   2. Stems very irregularly branched, ascending; branches julaceus, frequently themselves branched; leaf apex shortly apiculate-acuminate; plants in loose clumps, mainly on soil and rock in dry sites — 3. T. LESCURII

Plants green to brownish green, usually in tight mats; stems creeping, closely appressed, elongated, regularly pinnately branched; branches short, ascending, terete; leaves broadly ovate, abruptly apiculate-acuminate, very concave, margins stoutly toothed to ciliate; median cells each with a stout, curved, simple papilla; costa ending above midleaf. Sporophytes infrequently produced.

1. **Thelia hirtella** (Hedw.) Sull. Figure 46, E–F

  Habitat: Bark of trees and shrubs, occasionally on rock; sometimes on tree bases, logs, and stumps; forests.
  Range: Eastern North America; northeastern Mexico.
  Gulf Coast Distribution: Florida to Texas.

Usually this species is very easy to recognize in the field by its elongated, appressed, featherlike stems creeping on bark; it sometimes forms very dense mats but is often found as scattered stems, sometimes greatly elongated. Under the microscope its simple papillae distinguish it at once from *T. asprella* and *T. lescurii*.

I once found an unfinished nest of the Carolina wren that was almost entirely constructed of stems of *Thelia hirtella*. The nest was built in a cavity in the tongue of a small utility trailer, and the birds probably gave up on the construction because the space in the tongue was too small. The total weight of the nest was 7.8 grams, of which *Thelia hirtella* stems comprised 6.5 grams. The remainder of the nest material was composed of a few leaf blade fragments, leaf stalks, bits of grasses, a few twigs and roots, and bits of insect or spider cocoons.

In the forest area where I found the nest, *Thelia hirtella* is not common or abundant. Thus the wrens had probably searched a large area of the forest in order to selectively harvest this preferred moss, the only moss used in the nest. Other Carolina wren nests I have examined from the same general area contained mixtures of the mosses *Eurhynchium hians*, *Cryphaea glomerata*, *Thuidium delicatulum*, and *Forsstroemia trichomitria*, all of which are fairly common and abundant locally, as well as *Thelia hirtella*. Bent

(1948) cites mosses as being acceptable to Carolina wrens in nest construction but does not mention any specific kinds of mosses. Breil and Moyle (1976) recently studied the mosses used in bird nest construction at a site in Virginia.

2. Thelia asprella Sull.
Figure 46, G

Similar in general aspect to *T. hirtella*; plants more slender, in denser mats, more compact, branching less regularly pinnate; color gray-green, somewhat glaucous; leaves triangular, abruptly slenderly apiculate-acuminate, apiculation often spreading, margins ciliate, the cilia often branched and multicellular; median cells each with a stout, branched, dorsal papilla; costa ending about midleaf or a little above. Infrequently fruiting.

> **Habitat:** Mainly on bases of deciduous trees; stumps; occasionally on rocks; forests.
> **Range:** Eastern North America.
> **Gulf Coast Distribution:** Florida, Mississippi, and Texas.

The tree base habitat, dense mats of slender, pinnately branched plants, and the glaucous, gray-green color, usually allow this plant to be identified in the field. Under the microscope, the branched papillae separate it immediately from the similar *T. hirtella*; its compact growth habit, pinnate branching, slender stems and branches with the leaf tips spreading, and the filiform apiculation should allow it to be separated from *T. lescurii*. This species is much less common along the Gulf Coast than *T. hirtella* and *T. lescurii*. It should be looked for in interior areas, away from the coast. *Thelia asprella* has not yet been reported from Alabama or Louisiana but surely is present in these areas.

3. Thelia lescurii Sull.
Figure 46, H–I

Plants glaucous, gray-brown, in loose or dense, sometimes straggly mats; stems very irregularly branched with the branches often forked and of uneven length, both stems and branches ascending, julaceus; leaves very concave, closely imbricate, tips shortly apiculate, appressed, margins toothed and often ciliate; median cells of leaves each with a stout, branched, dorsal papilla; costa ending at or above midleaf. Infrequently fruiting.

> **Habitat:** Mainly on soil, sometimes on rocks and tree bases, usually in dry, rather open sites with sandy soil.
> **Range:** Eastern North America.
> **Gulf Coast Distribution:** Florida to Texas.

Due to its straggly habit with julaceus stems and branches, irregular branching, and pale, glaucous appearance, *T. lescurii* does not really resemble either *T. asprella* or *T. hirtella* in the field. Although it sometimes forms dense mats, still the ascending stems and branches give it a quite different aspect than either of the other species of *Thelia*. It is most often found in dry, sandy, thinly wooded sites, rather than in dense forests.

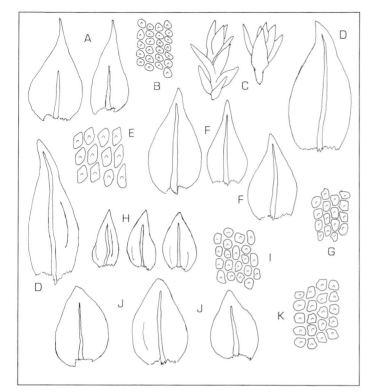

Fig. 47. *A–C:* **Lindbergia bra-chyptera.** *A*, leaves (× 27). *B*, cells from lower portion of leaf (× 231). *C*, deciduous branchlets from upper leaf axils (× 55). *D–E:* **Leskea polycarpa.** *D*, leaves (× 55). *E*, median cells (× 385). *F–G:* **Leskea australis.** *F*, leaves (× 55). *G*, median cells (× 385). *H–I:* **Leskea gracilescens.** *H*, leaves (× 27). *I*, median cells (× 385). *J–K:* **Leskea obscura.** *J*, leaves (× 26). *K*, median cells (× 385).

Dull, brownish green plants in thin mats or as scattered stems; stems creeping, irregularly branched, paraphyllia few or absent; branches short, spreading, frequently bearing clusters of small, leafy propagula in axils of upper leaves; leaves appressed when dry, squarrose when moist, ovate-acuminate, apical cells elongate and ± hyaline, margins plane, entire to slightly serrulate; median cells thick-walled, isodiametric, with rounded corners, bluntly uni-papillose on both surfaces, mostly in distinct vertical rows, especially in lower part of leaf; costa ending about midleaf or a little above. Capsules ± erect and symmetric; endostome a low membrane; operculum conic.

> **Habitat:** Tree trunks, especially those with roughly textured bark such as sweet gum and post oak; in open, rather dry forests.
>
> **Range:** Eastern North America; Arizona and western Texas; Japan; Caucasus and Himalayas.
>
> **Gulf Coast Distribution:** Known in the Gulf Coast only from two collections by P. L. Redfearn in central Louisiana, Natchitoches Parish, in 1976.

The leaves of this moss are squarrose when wet, and this feature, together with the slender, hyaline leaf tip, papillose cells, and the axillary propagula, makes the species distinctive. Its very

2. LINDBERGIA Kindb.

Lindbergia brachyptera (Mitt.) Kindb.
Figure 47, A–C

small size and inconspicuous habit make it difficult to find in the field, where it may resemble the creeping stems of *Cryphaea*, *Platygyrium*, *Leucodon*, or even *Forsstroemia*. However, its microscopic appearance is distinctive. I have no doubt that *L. brachyptera* occurs in eastern Texas, where it should be searched for in post oak areas, and it very likely occurs as well in parts of Alabama and Mississippi.

3. LESKEA Hedw.

Plants dull green to brownish green, in thin, often straggly mats; stems usually irregularly branched, sometimes pinnate; branches short or elongate; paraphyllia scanty, inconspicuous, attenuate to narrowly lanceolate; leaves appressed when dry, concave, ovate-triangular, shortly acuminate, broadly acute, or obtuse, apex often oblique; cells small, isodiametric, obscure, bluntly unipapillose on one or both surfaces; margins entire or ± serrulate above; costa strong, ending in acumen. Capsules erect and symmetric or curved; endostome sometimes reduced; operculum conic.

Leskea is a small genus of mostly small, rather inconspicuous mosses growing in our area mainly on tree bases and roots in moist sites. Plants of the genus can usually be easily recognized by the habitat and habit, papillose cells, irregular branching, and the usually erect and symmetric capsules. The paraphyllia, although usually present, are often difficult to find because so few in number. Four species of *Leskea* occur along the Gulf Coast. One of them, *L. australis*, is fairly common and sometimes abundant; the others are infrequently encountered and rarely abundant. All four species are frequently found with sporophytes in our area.

1. Capsules curved; leaf tips acuminate, usually oblique; leaves often secund	1. L. POLYCARPA
1. Capsules erect, straight; leaf tips acuminate or obtuse or broadly acute, oblique or straight; leaves not secund	2
2. Plants very small and slender; leaf tips acuminate; stems and branches uniformly slender, all about the same diameter	2. L. AUSTRALIS
2. Plants coarser; leaf tips broadly acute or obtuse, not acuminate; stems and branches usually of different diameters, some conspicuously thicker than others	3
3. Leaf apex usually very obtuse or broadly acute; leaves not usually appearing plicate below; lower margins not noticeably revolute	4. L. OBSCURA
3. Leaf apex usually acute; leaves often appearing biplicate below; lower margins often ± revolute	3. L. GRACILESCENS

1. Leskea polycarpa Hedw.
Figure 47, D–E

Dark green to brownish green plants in thin mats; stems pinnately branched; branches spreading; leaves often ± secund at branch tips, ovate-lanceolate, rather abruptly acuminate with acumination usually oblique, concave, lower margins usually ± recurved, leaf appearing biplicate; costa ending in acumen. Capsules cylindric, curved.

> **Habitat:** Tree bases, roots; wet forests subject to flooding.
> **Range:** Eastern North America; Europe; Asia; Africa.
> **Gulf Coast Distribution:** Louisiana; eastern Texas. Apparently not yet reported from Florida, Alabama, or Mississippi, but to be expected there. It is neither common nor abundant in our area.

With its curved capsules and often secund leaves, *L. polycarpa* is one of the most easily identified species of *Leskea*.

2. Leskea australis Sharp
Figure 47, F–G

Plants in thin, dark green to brownish green or blackish mats; stems and branches very slender, terete when dry; branches ascending-spreading, elongated, tapering; leaves small, 0.5 mm long or less, ovate-lanceolate, short-acuminate, apex acute or obtuse, sometimes oblique, margins plane; costa strong, ending in acumen. Capsules erect, straight.

> **Habitat:** Tree trunks, bases and roots; logs and soil; in low forests, usually subject to flooding.
> **Range:** Southeastern United States.
> **Gulf Coast Distribution:** Florida to Texas.

Leskea australis is a well-marked species and is not easily confused with the other three species that occur in our area. The very slender branches bearing tiny leaves are characteristic, and the quality of having the branches and stems of uniform diameter rather than of differing sizes is also helpful in recognition of *L. australis*. This species is rather common throughout our area, from Florida well into eastern Texas.

3. Leskea gracilescens Hedw.
Figure 47, H–I

Plants in thin, dark green to brownish green mats; stems irregularly pinnately branched; branches spreading, elongate; leaves basically ovate, concave, apex acute or broadly acute, often oblique; leaf base often appearing biplicate; margins ± recurved below; costa ending in acumen. Capsules erect and symmetric.

> **Habitat:** Tree trunks and bases, roots, rocks; forests.
> **Range:** Eastern North America.
> **Gulf Coast Distribution:** Florida to eastern Texas.

It is sometimes difficult to decide whether a given specimen of *Leskea* fits best within the concept of *L. gracilescens* or in that of *L. obscura*, but most specimens can be determined fairly conveniently. The generally acute leaf tips, often recurved lower margins, and the biplicate appearance of the concave leaves will usually serve to distinguish *L. gracilescens* from *L. obscura*.

4. Leskea obscura Hedw.
Figure 47, J–K

Similar to *L. gracilescens* in many respects; differing in the usually nonplicate leaves with erect margins and generally much more obtuse tips. The leaves are also more broadly ovate and mostly lack a definite acumination. The apex of the leaf in both species is often oblique.

> **Habitat:** Tree trunks and bases, logs, roots, soil, rocks; low forests.
> **Range:** Eastern North America; Europe; Asia.
> **Gulf Coast Distribution:** Florida to eastern Texas.

Probably more common than *L. gracilescens* in our area and occupying essentially the same habitats. See remarks under *L. gracilescens*.

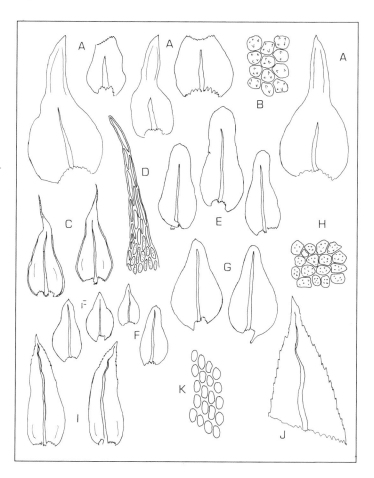

Fig. 48. *A–B:* **Haplohymenium triste.** *A,* leaves, two with tips characteristically broken off (× 55). *B,* median cells (× 385). *C–D:* **Anomodon rostratus.** *C,* leaves (× 3). *D,* leaf tip (× 247). *E,* **Anomodon minor,** branch leaves (× 14). *F–H:* **Anomodon attenuatus.** *F,* branch leaves (× 14). *G,* stem leaves (× 14). *H,* median cells of branch leaf (× 412). *I–K:* **Herpetineuron toccoae.** *I,* leaves (× 14). *J,* leaf tip (× 55). *K,* median cells (× 385).

4. HAPLOHYMENIUM
Dozy & Molk.

Haplohymenium triste (Ces. ex De Not.) Kindb.
Figure 48, A–B

Plants slender, in thin, brownish green, straggly mats; stems creeping, irregularly branched, without paraphyllia; branches spreading or ascending, slender, terete, simple; stem and branch leaves similar; leaves mostly with ovate to broadly ovate base and acuminate to lingulate tip, apex blunt or acute, often with a sharp, elongated, papillose terminal cell; leaf tips usually broken off in older portions of stems and branches; median leaf cells conspicuously bulging-unipapillose or pluripapillose, isodiametric; leaf margins serrate-dentate by bulging-papillose cells; costa ending about midleaf. Not fruiting in our area.

Habitat: Trunks and logs of broadleaf trees in forests; especially common on *Fagus grandifolia* and *Magnolia grandiflora*.
Range: Eastern North America; Mexico; Europe; Asia; Hawaii.
Gulf Coast Distribution: Florida to Texas.

Haplohymenium triste is rather uncommon in our area, and because of its slender stems and branches and inconspicuous habit, it is not easy to find. However, careful search on the trunks of large broadleaf trees in forested areas will often turn it up. Its general appearance in the field is suggestive of a slender *Anomodon*, but its uniformly slender stems and branches, with the leaf tips broken off in older portions, identify it easily.

5. ANOMODON Hook. & Tayl.

Dull, dark green to brownish or yellowish green plants in tight or loose mats; primary stems creeping, their leaves sometimes much reduced, bearing numerous, mostly ± erect or curved branches that are often themselves branched; paraphyllia lacking; leaves basically with ovate base and lingulate to acuminate tip; median cells small, opaque, pluripapillose, isodiametric; costa strong, mostly ending well below leaf apex. Capsules erect and symmetric; peristome pale; operculum rostrate.

Three species of *Anomodon* occur in our area; one of them, *A. attenuatus*, is very common, and the other two are less frequently encountered. The three are very easily distinguished from one another, even in the field.

1. Stems and branches terete; plants yellowish green; leaves with long, hyaline hair-points, margins revolute	1. A. ROSTRATUS
1. Stems and branches often somewhat flattened when dry; plants dark or brownish green; leaves rounded or apiculate at tips, lacking hair-points, margins plane	2
2. Leaf tips lingulate, apex broadly rounded	2. A. MINOR
2. Leaf tips mostly acute to apiculate	3. A. ATTENUATUS

Plants in compact, usually yellowish green mats; stems creeping, rather regularly pinnately branching; branches ascending, short or elongate; stems and branches terete when dry, not flattened; leaves with ovate base and tapering tip, apex acute, usually with a hair-point of narrow, elongate cells, point sometimes reduced, margins revolute; costa strong, ending in acumen.

1. Anomodon rostratus
(Hedw.) Schimp.
Figure 48, C–D

Habitat: Tree bases, soil, rock, usually in well-drained sites in forests.

Range: North America, mostly east of the Rockies; Europe; Mexico; Guatemala; Bermuda; Jamaica; Haiti.
Gulf Coast Distribution: Florida to Texas.

Anomodon rostratus occurs virtually throughout the Gulf Coast in forested areas, but it is not as common as *A. attenuatus*. It can be easily recognized in the field by its color, compact habit, and the terete branches bearing long-pointed leaves. Its aspect in the field is quite different from that of the other two species of *Anomodon* in our area; it more resembles a *Thuidium*, perhaps.

2. **Anomodon minor** (Hedw.) Fürnr.
Figure 48, E

Plants dull, green to brownish green, in usually loose, often thin mats; stems creeping; branches elongate, mostly simple, ascending, straight or curved; leaves with ovate base and lingulate tip, apex broadly rounded, not apiculate, tips of older leaves mostly broken off; costa strong, ending well below leaf apex.

> **Habitat:** Trees, especially on the bases; rocks; forests.
> **Range:** North America, mostly east of the Rockies; Asia; Europe; Mexico.
> **Gulf Coast Distribution:** Florida to Texas.

This distinctive species is much less common in our area than is *A. attenuatus* but is just as widely distributed along the Gulf Coast. It shares essentially the same types of habitats as does *A. attenuatus*, although probably in sites a little drier. The two species are easily distinguished by the lingulate leaf tips of *A. minor*, which are broadly rounded at the apex. When dry, the flat leaf tips overlap one another like shingles, giving a characteristic appearance to the plants in the field.

3. **Anomodon attenuatus** (Hedw.) Hüb.
Figure 48, F–H

Plants dark green to brownish green, dull, in intricate, usually dense mats; stems and branches rather coarse; major branches usually curved downwards, ± flattened, usually many branches elongate and attenuate at ends; leaves with ovate to broadly ovate bases and tapering tips, apex often with a sharp apiculation; costa strong, ending in acumination.

> **Habitat:** Tree trunks and bases; rocks; soil; ravines and banks in forests; everywhere.
> **Range:** North America, mostly east of the Rockies; Europe; Asia; Mexico; Guatemala; Cuba; Jamaica.
> **Gulf Coast Distribution:** Florida to Texas.

This is almost a weedy species in the sense that it occurs pretty much throughout our area and is often abundant. The dull mats that it forms on tree bases are characteristic, and the species can thus be recognized even from a distance of several feet. The flattening of stems and branches is more pronounced in some specimens than in others and may even be so evident that beginning students of mosses have taken it for a *Fissidens* in the field. The tapering, often apiculate leaf apex distinguishes *A. attenuatus* under the micro-scope from *A. minor*, in which the leaf tip is broadly lingulate and rounded. These qualities can also be seen in the field with a hand lens. The presence of many attenuate branches is also diagnostic for *A. attenuatus*.

Plants in thin, dark green to brownish, often straggly mats; primary stems stoloniform, very slender, nearly naked or with small, highly reduced leaves; branches ascending, mostly simple, ± circinately coiled when dry; flagelliform, microphyllous branches common, sometimes abundant, arising from stems or tips of regular branches, fragile and easily dehiscing; branch leaves lanceolate to ovate-lanceolate, concave and ± biplicate at base, apex often apiculate, leaves often asymmetric; median cells isodiametric, smooth, mostly in regular vertical rows and forming oblique horizontal rows be-tween costa and leaf margins; margins plane above, somewhat re-curved at base, usually coarsely serrate above but sometimes al-most entire; costa strong, serpentine in upper portion, ending in acumen. Not fruiting in our area.

6. HERPETINEURON (C.M.) Card.

Herpetineuron toccoae (Sull. & Lesq. ex Sull.) Card.
Figure 48, I–K

> **Habitat:** Tree trunks, rocks, logs; rich deciduous forests.
> **Range:** Southeastern United States; western Texas; Mexico; Guatemala; South America; Asia; Africa; East Indies.
> **Gulf Coast Distribution:** Florida to Louisiana. Not yet re-ported from the Panhandle counties of Florida or from east-ern Texas, but probably present in both areas.

Herpetineuron toccoae is not common in our area and is rarely abundant. It is easily recognized in the field by the circinate, mostly simple branches. The leaves are ± involute when dry, giving the branches a characteristic braided appearance. The flagelliform branches also help distinguish this species in the field. Under the microscope the sinuous costa, serrate margins, and small, smooth cells are helpful in recognition. In the absence of spores, it is likely that the easily detached flagelliform branches serve for propagation of this moss.

THUIDIACEAE Schimp.

Plants small to medium-size or robust, mostly dull, soft to firm or wirey, yellowish to dark green or brownish, in thin, spreading colo-nies or dense, springy mats; paraphyllia mostly present, often abun-dant and conspicuous; stems prostrate to arched-ascending, freely and often regularly pinnately branched; branch leaves often dif-ferentiated from stem leaves; leaves mostly concave and ovate, apex blunt to acuminate; cells uni- to pluripapillose, mostly ± isodiametric; costa usually strong and single; perichaetial leaves differentiated. Setae elongate; capsules inclined and asymmetric; peristome double; operculum convex to rostrate; calyptra cucullate, naked.

This family is represented in our area by two genera, *Thuidium* and *Haplocladium*; other genera occur in North America, to the north. The strongly curved (when mature and dry) capsules and often abundant paraphyllia help differentiate the family from the preceding one, the Leskeaceae.

Apical cell of branch leaves with a single terminal papilla; plants
   once pinnately branched, not fernlike in appearance; paraphyllia
   mostly neither abundant nor conspicuous   1. HAPLOCLADIUM

Apical cell of branch leaves with two to several terminal papillae;
   plants mostly ± regularly once to three times pinnately branched,
   often fernlike; paraphyllia very abundant and conspicuous   2. THUIDIUM

---

1. HAPLOCLADIUM (C.M.) C.M.

Haplocladium microphyllum
(Hedw.) Broth.
Figure 49, A–B

Plants in dull, usually thin, yellowish to brownish green mats; stems pinnately branched, often irregularly so, bearing small, linear or branched paraphyllia; branches spreading or ascending, short or somewhat elongate, simple; stem leaves with long acumen from concave, broadly ovate, somewhat biplicate base, margins ± recurved; branch leaves smaller, ovate-lanceolate, apex acute, median cells each with a single, low papilla, papillae sometimes very low and difficult to see, occasionally apparently lacking; costa strong, ending in acumen. Frequently fruiting; capsules small and strongly curved when dry and empty.

> **Habitat:** Soil, logs, stumps, tree bases, rotted wood; wet forests and damp shady sites in general, including lawns.
> **Range:** North America; Central and South America; West Indies; Asia; Europe.
> **Gulf Coast Distribution:** Florida to Texas.

This is one of the most common species of Gulf Coast mosses; it is quite weedy and, because it is variable in general appearance, it takes some practice to recognize it consistently in the field. In some respects *H. microphyllum* resembles species of *Amblystegium*, and it often grows in similar habitats. However, the papillose leaf cells and the paraphyllia identify it under the microscope.

HAPLOCLADIUM VIRGINIANUM (Brid.) Broth. has been reported from our area in Alabama and Texas; however, I have not seen any authentic specimens and doubt that it actually occurs in the Gulf Coast area. It differs from *H. microphyllum* in having abruptly acuminate stem leaves and in the plants being rigid with crowded leaves. Further, the stem leaves are not plicate and have the margins erect, rather than ± recurved as in *H. microphyllum*. In general appearance *H. virginianum* more closely resembles *Thuidium*

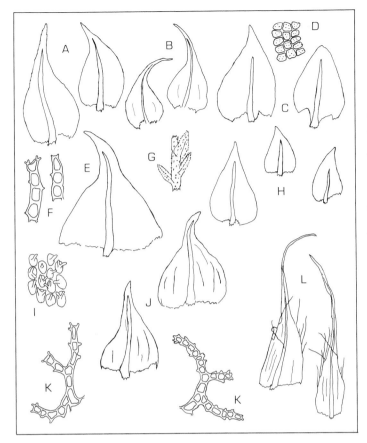

Fig. 49. *A–B*: **Haplocladium microphyllum.** *A*, branch leaves (× 15). *B*, stem leaves (× 27). *C–F*: **Thuidium minutulum.** *C*, branch leaves (× 88). *D*, median cells of branch leaf (× 340). *E*, stem leaf (× 88). *F*, paraphyllia (× 340). *G*, **Thuidium pygmaeum,** portion of branch with leaves (× 55). *H–L*: **Thuidium delicatulum.** *H*, branch leaves (× 27). *I*, median cells of branch leaf (× 436). *J*, stem leaves (× 27). *K*, paraphyllia (× 231). *L*, perichaetial leaves (× 15).

minutulum than *Haplocladium microphyllum*; it differs from the former, among other ways, in having only one papilla per cell.

Plants slender to rather robust, dull, green to yellowish or brownish, in loose, often springy mats or cushions or forming thin mats; stems once to three times pinnately branched, prostrate and creeping or ascending-arched, with sparse or abundant paraphyllia; branches spreading, paraphyllia usually few; stem and branch leaves differentiated; leaf cells isodiametric, uni- or pluripapillose; costa strong, ending about midleaf or in acumen; paraphyllia filamentous and simple or branched, to linear and several cells wide, usually papillose. Capsules asymmetric; peristome double; operculum conic to long-rostrate.

Three species of *Thuidium* are common along the Gulf Coast, especially in the central and eastern portions. A fourth species has been found in the Florida Panhandle, at the eastern limit of our range. The dull appearance of the plants, and the usually regularly pinnate habit, make the genus easy to recognize. In general, these are plants of woodland habitats, growing on soil, logs, humus and tree bases in rich mesic forests.

2. THUIDIUM B.S.G.

1. Plants slender, once or twice pinnately branched; paraphyllia filamentous, unbranched; median cells of leaves pluripapillose   2
1. Plants medium-size to robust, twice to three times pinnately branched; paraphyllia multiform, frequently branched; median cells of leaves uni- or pluripapillose   3
    2. Stems and branches smooth (other than the paraphyllia); on old wood and soil   1. T. MINUTULUM
    2. Stems and branches papillose; on rock, especially limestone   2. T. PYGMAEUM
3. Plants regularly twice to three times pinnately branched, usually symmetric in outline; perichaetial leaves ciliate; median cells of branch leaves each with a large, often forked papilla; stem leaves strongly papillose, with long, slender acumination   3. T. DELICATULUM
3. Plants irregularly twice pinnately branched (rarely three times pinnately branched here and there), straggly in appearance; perichaetial leaves lacking cilia; median cells of branch leaves each with one to several low papillae; stem leaves inconspicuously papillose, with short acumination   4. T. ALLENII

---

1. **Thuidium minutulum**
(Hedw.) B.S.G.
Figure 49, C–F

Plants slender, dull, dark green to brownish or yellowish, in thin or springy intricate mats; stems prostrate with few to abundant paraphyllia, mostly once pinnately branched; branches often themselves irregularly branched, spreading, short, with few paraphyllia; paraphyllia short, filamentous, simple, papillose; stem leaves incurved-appressed when dry, deltoid-acuminate to broadly ovate-acuminate, concave, biplicate, margins recurved, median cells pluripapillose on both surfaces, terminal cell often elongate and with several papillae on its tip, costa ending in acumen; branch leaves strongly incurved when dry, spreading-ascending when moist, much smaller than stem leaves, ovate-lanceolate, slightly concave, margins plane to slightly recurved, beset with projecting-papillose cells, median cells mostly pluripapillose, some unipapillose, terminal cell crowned with several papillae, costa ending just above midleaf or in base of acumen; perichaetial leaves lacking marginal cilia. Capsules very asymmetric, inclined to horizontal; operculum rostrate.

Habitat: Soil, rocks, rotted logs, roots, tree bases; deep forests.
Range: Eastern North America; tropical and South America; Europe; Bermuda; Azores.
Gulf Coast Distribution: Florida to eastern Texas.

Thuidium minutulum is easy to recognize because of its slender form, mostly once-pinnate stems, and the strongly incurved-appressed branch leaves when dry. Under the microscope the short, unbranched paraphyllia are distinctive. Although the stems of this moss are once pinnately branched, with the branches themselves only irregularly branched, some plants may be mostly twice pinnately branched and thus resemble superficially *T. delicatulum* or *T. allenii*. However, the simple paraphyllia and the mostly uni-papillose cells should differentiate *T. minutulum* easily.

Similar to *T. minutulum* but smaller and more slender, and with the stems and branches papillose. In both *T. minutulum* and *T. pygmaeum* the stems and branches bear paraphyllia, but in the latter there are papillae also.

2. Thuidium pygmaeum
B.S.G.
Figure 49, G

> **Habitat:** Rock exposures, limestone and sandstone in forests.
> **Range:** Eastern North America; Asia.
> **Gulf Coast Distribution:** Florida.

This little moss has been found in our area only in Jackson County, Florida, where it grew on limestone at Florida Caverns State Park. It could well be expected to turn up in the western part of our range. It much resembles *T. minutulum*, but the plants are considerably smaller. The papillae on the stems and branches are distinctive.

Plants robust, green to yellowish or brownish green, in dense, intricate, springy, often extensive mats or cushions; stems prostrate to arched-ascending, bearing abundant paraphyllia, mostly regularly three times pinnately branched and fernlike in appearance; branches spreading; paraphyllia multiform, papillose, filamentous and simple to branched or several cells wide; stem leaves incurved when dry, concave, deltoid- to ovate-acuminate, often 2–4 plicate, sometimes strongly so, margins recurved, median cells each with a stout, sometimes forked papilla on each surface, terminal cell elongated or short, with several apical papillae, costa ending in middle of acumen; branch leaves imbricate and somewhat incurved when dry, ascending when moist, short-ovate-acuminate, concave, margins erect, armed with projecting, sharply papillose cells, median cells each with a stout, often forked and inclined dorsal papilla, ventral papillae smaller, terminal cells crowned with several papillae, costa ending in base of acumen; inner perichaetial leaves plicate, long and slender, with remarkably elongate tips, bearing long, delicate marginal cilia. Setae very long; capsules elongate, asymmetric, inclined; operculum long and often obliquely rostrate.

3. Thuidium delicatulum
(Hedw.) B.S.G.
Figure 49, H–L

> **Habitat:** Soil, rocks, logs, tree bases; ravines, stream banks in rich forests.
> **Range:** Widespread in northern and eastern North America; tropical and South America; Europe; Asia.
> **Gulf Coast Distribution:** Florida to Texas.

Thuidium delicatulum is a beautiful and rather common moss. Its fernlike aspect and conspicuous size make it easy to identify; it is often referred to as "fern moss," making it one of the few mosses to have a common name that is really useful. It is a favorite for use in terrariums or winter gardens, where it proves long-lived and durable, although eventually usually becoming overgrown with molds and algae, a common fate of mosses in terrariums.

From *T. allenii*, *T. delicatulum* differs most obviously in its regular branching and symmetric appearance, in the conspicuously ciliate perichaetial leaves, in the more strongly papillose cells of the branch leaves, and in the shape of the stem leaves. Its very regularly twice to three times pinnately branching and symmetric form make it easy to recognize in the field.

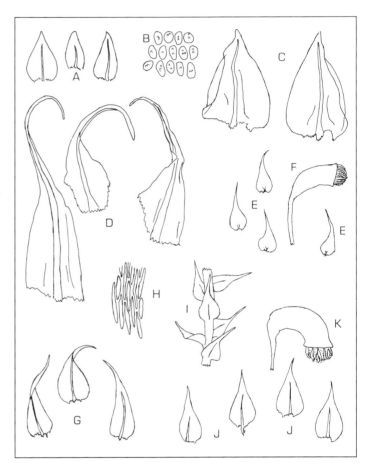

Fig. 50. *A–D*: **Thuidium allenii.** *A*, branch leaves (× 27). *B*, median cells of branch leaf (× 412). *C*, stem leaves (× 24). *D*, perichaetial leaves (× 22). *E–F*: **Campylium hispidulum.** *E*, leaves (× 15). *F*, capsule (× 15). *G–H*: **Campylium chrysophyllum.** *G*, leaves (× 15). *H*, median cells (× 231). *I–K*: **Campylium radicale.** *I*, portion of branch with leaves (× 15). *J*, leaves (× 15). *K*, capsule (× 15).

4. **Thuidium allenii** Aust.
Figure 50, A–D

Similar in some respects to *T. delicatulum* but differing as stated in the key. Another difference is that the stem leaves of *T. allenii* have the margins plane to erect or even slightly recurved, whereas those of *T. delicatulum* have the margins strongly recurved. In addition, the stem leaves of *T. allenii* are rarely plicate but those of *T. delicatulum* are usually strongly and conspicuously biplicate. When the plants are fruiting, the absence of long, delicate cilia on the margins of the inner perichaetial leaves is distinctive for *T. allenii*. (Be careful not to mistake the long, delicate paraphyses, usually present in the perichaetium, for cilia.)

Habitat: Soil, logs, tree bases, humus, cypress knees; low forests. Generally in moister sites than *T. delicatulum*.

Range: Eastern United States.

Gulf Coast Distribution: Florida to eastern Texas. Not yet reported from the Panhandle counties of Florida but surely present there.

Thuidium allenii can usually be recognized in the field by its straggly appearance, as opposed to the neat fernlike aspect of typical *T. delicatulum*. It is probably much more common and widely distributed than existing records indicate, since it is quite likely to be passed over in the field as a poorly developed form of *T. delicatulum*. This moss only rarely produces sporophytes; in the single fruiting specimen I have seen, the capsules are immature, strongly curved, and the operculum is conic. I have seen a few other specimens with old setae and also several with perichaetia containing archegonia.

THUIDIUM RECOGNITUM (Hedw.) Lindb. has not been reported from the Gulf Coast but could occur in inland areas. It resembles *T. delicatulum* but differs in, for example, lacking cilia on the perichaetial leaves, the paraphyllia papillose at or near the upper ends of the cells rather than with scattered papillae, and the operculum conic rather than long-rostrate. From *T. allenii*, *T. recognitum* can be distinguished by the stoutly unipapillose cells of its branch leaves and by its plicate, abruptly acuminate stem leaves.

AMBLYSTEGIACEAE (Broth.) Fleisch.

Plants small to robust, mostly yellowish green, in thin to dense mats; stems prostrate and creeping to erect-ascending, mostly not flattened, usually freely and irregularly branched, sometimes subpinnately to pinnately branched; leaves mostly small and lanceolate-acuminate; cells oblong to linear, usually smooth, often differentiated in alar regions; costa single, short and double, or lacking. Setae elongate; capsules cylindric-curved; peristome double; operculum convex to conic or apiculate; calyptra cucullate.

The Amblystegiaceae are primarily plants of wet places, although few are actually aquatic and some occupy fairly dry habitats. The rather slender, curved capsules of the Amblystegiaceae help distinguish them from the Brachytheciaceae, which typically have shorter, blunt capsules and which typically occur in drier habitats. The peristome teeth of the Amblystegiaceae are yellow-brown, in contrast to the red-brown teeth of the Brachytheciaceae.

1. Leaves conspicuously bordered with elongate, thick-walled cells — 2. SCIAROMIUM
1. Leaves not bordered — 2
   2. Costa single or double and short; leaves firm, mostly squarrose or wide-spreading above the base, with channeled acumen; plants mostly of rather dry sites — 1. CAMPYLIUM
   2. Costa single; leaves mostly soft, erect to spreading from insertion, acumen plane; plants of mostly rather wet sites — 3. AMBLYSTEGIUM

1. CAMPYLIUM (Sull.) Mitt.

Plants small to medium-size, green to yellowish or brownish green, often glossy, in thin to dense mats; stems creeping or ascending, freely and mostly irregularly branched; leaves nearly erect to wide-spreading or squarrose, sometimes falcate-secund, mostly slenderly acuminate from a broad base, margins mostly entire; cells rhombic to linear, smooth; alar cells in ± distinct groups, subquadrate and dense to enlarged and inflated; costa lacking, short and double, or single and reaching midleaf or beyond. Capsules curved; operculum pointed.

Four species of *Campylium* occur on the Gulf Coast. One species, *C. chrysophyllum*, is very common and often abundant over much of our area. *Campylium* can be recognized, in a general way, by the color and small size of the plants and by the spreading to squarrose leaves.

1. Leaves ecostate or costa very short and double — 1. C. HISPIDULUM
1. Leaves with single, well-developed costa — 2
   2. Leaves erect-spreading, 2–2.5 mm long; alar cells conspicuous, ± enlarged and inflated — 4. C. POLYGAMUM
   2. Leaves spreading to wide-spreading or squarrose, less than 2 mm long; alar cells not particularly inflated or conspicuous — 3
3. Leaves crowded on stems, often ± falcate-secund at stem and branch tips; alar cells small, dense — 2. C. CHRYSOPHYLLUM
3. Leaves distant on stem, wide-spreading, not falcate-secund; alar cells slightly enlarged, not dense — 3. C. RADICALE

1. Campylium hispidulum
(Brid.) Mitt.
Figure 50, E–F

Plants very small, yellowish green to green, in thin, creeping mats; stems prostrate to ascending, irregularly to subpinnately branched, firmly attached to substrate; leaves small, about 0.5–0.75 mm long, abruptly slenderly acuminate from broadly ovate base, strongly squarrose, crowded on stem, margins serrulate; cells oblong to ± fusiform or oblong-linear, quadrate in basal angles; costa lacking or very short and double.

Habitat: Tree and shrub bases, rotted logs, stumps, soil and rock; in rather dry to mesic forests.
Range: Widespread in the Northern Hemisphere.
Gulf Coast Distribution: Florida to Texas.

This little moss is easy to recognize due to its small size, squarrose leaves, and lack of a costa. It is fairly common in our area in upland sites.

Plants small to medium-size, yellowish to brownish green, ± glossy, in thin to dense mats; stems creeping to ascending, highly and irregularly branched; leaves crowded, spreading to ± squarrose, gradually or abruptly slenderly acuminate from an ovate to triangular base, commonly falcate and secund, entire or slightly serrulate at base; cells oblong-linear to linear; alar cells small, dense, subquadrate, thick-walled, forming distinct groups; costa extending to midleaf or beyond.

2. Campylium chrysophyllum
(Brid.) J. Lange
Figure 50, G–H

> **Habitat:** Tree bases, ravine and stream banks, rocks; mesic to
> dry forests; more common in drier sites.
> **Range:** Widespread in the Northern Hemisphere; tropical
> America.
> **Gulf Coast Distribution:** Florida to Texas.

This moss is exceedingly common and abundant over most of our area. Most of the Gulf Coast specimens have the leaves somewhat to strongly falcate-secund, making them easy to recognize. The conspicuously falcate-secund plants have been named var. *brevifolium* (Ren. & Card.) Grout. Plants of *C. radicale* are generally smaller, have usually shorter upper cells, and have leaves widely spaced on the stems.

Plants small, yellowish green, in thin, spreading mats; stems creeping or ascending, irregularly branched; leaves distant on stems, wide-spreading, abruptly slenderly acuminate from ovate base, 1–1.3 mm long, entire or slightly serrulate; cells oblong-rhombic to linear, enlarged in basal angles; costa single, extending to midleaf or into base of acumen.

3. Campylium radicale
(P.-Beauv.) Grout
Figure 50, I–K

> **Habitat:** Stream banks, around springs; soil and humus; low
> forests.
> **Range:** Europe; eastern North America.
> **Gulf Coast Distribution:** Mississippi and Louisiana.

Campylium radicale is not at all common in our area, at least judging from extant specimens, but it may be overlooked by collectors due to its obscure habitat and inconspicuous nature. It is easily distinguished from *C. chrysophyllum* by its leaves, which are non-secund and distant from one another on the stem. This moss is somewhat like *Amblystegium* both in its habit and habitat; however, the long, channeled leaf tip and somewhat enlarged alar cells, as well as the squarrose leaves, distinguish it easily.

Plants glossy, golden brown to yellowish green, rather robust; leaves erect-spreading, 2–2.5 mm long, alar cells ± enlarged and inflated, forming conspicuous groups at the basal angles.

4. Campylium polygamum
(B.S.G.) C. Jens.
Figure 51, A

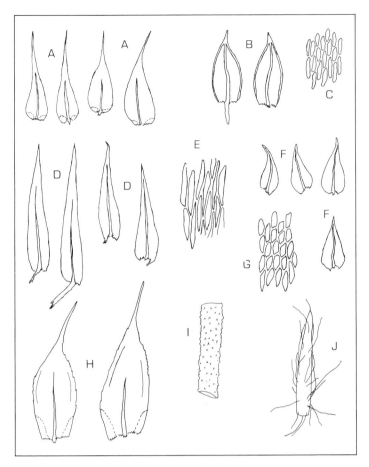

Fig. 51. *A,* **Campylium polyg-amum,** leaves (× 15). *B–C:* **Scia-romium lescurii.** *B,* leaves (× 15). *C,* median cells (× 231). *D–E:* **Am-blystegium riparium.** *D,* leaves (× 15). *E,* median cells (× 231). *F–G:* **Amblystegium varium.** *F,* leaves (× 15). *G,* median cells (× 231). *H–J:* **Homalotheciella subcapil-lata.** *H,* leaves (× 55). *I,* portion of seta (× 55). *J,* immature calyptra (× 15).

Habitat: Soil and humus in wet places.
Range: Europe; North America.
Gulf Coast Distribution: Florida.

This moss is known in our area only from Jackson County, Florida. It could well turn up, however, farther westward along the Gulf Coast in inland areas. The longer leaves, neither squar-rose nor secund, and the inflated alar cells distinguish it from *C. chrysophyllum.*

2. SCIAROMIUM (Mitt.) Mitt.

Sciaromium lescurii (Sull.) Broth.
Figure 51, B–C

Platylomella lescurii (Sull.) Andr.

Plants in loose or dense, sordid mats, blackish or brownish except for yellowish green growing tips of stems and branches; stems irreg-ularly branched, prostrate or ascending; leaves loosely erect, ± shriveled when dry, glossy, decurrent, ovate to ovate-lanceolate, acute to abruptly acuminate, strongly bordered all around by sev-eral rows of thick-walled, elongate cells, the border several cell layers thick and merging with the costa above to form the acumen; cells short-rhombic to fusiform or short-vermicular; costa strong, forming the acumen with the border. Capsules curved; operculum conic-apiculate.

Habitat: On rocks in cool, clear streams and springs; submerged or emergent.

Range: Eastern North America.

Gulf Coast Distribution: Louisiana.

This distinctive moss is known along the Gulf Coast only from a few collections in central Louisiana. It could well turn up, however, in other places in our area. The mats of plants are often encrusted with diatoms and other algae, except for the growing tips.

Plants small to rather robust, slender, soft to somewhat rigid, dull, green to yellowish or brownish, in thin to dense, sometimes floating, mats; stems prostrate, irregularly and usually freely branched; leaves ovate to lanceolate, obtuse to acuminate, entire to serrulate, sometimes ± complanate; cells short-rhombic to elongate-hexagonal or ± linear; costa extending to midleaf or beyond, sometimes ending in the acumen or excurrent. Capsules curved, constricted below the mouth when dry; operculum conic to conic-apiculate.

Plants of this genus are notoriously variable, as are many plants of aquatic and subaquatic habitats. They are often rather drab and inconspicuous. The genus is said to be difficult, meaning that it is not always easy to identify specimens at the specific level. Fortunately, however, only two species occur along the Gulf Coast, and they are easily distinguished from one another.

The genus *Leptodictyum* is often recognized as distinct from *Amblystegium*; it would include, in our area, *A. riparium*. But, as persuasively explained by Crum and Anderson (1981), there is little logic to separating *Leptodictyum* from *Amblystegium*. In addition to the two species of *Amblystegium* included here, several others have been reported from the Gulf Coast (some as *Leptodictyum*, one as *Hygroamblystegium*), but review of specimens shows clearly that only *A. riparium* and *A. varium* actually occur here. Both are widespread and rather common, especially in the eastern part of our area.

3. AMBLYSTEGIUM B.S.G.

Leaves 2.5 mm or more long	1. A. RIPARIUM
Leaves smaller, to 1.5 mm long, mostly shorter	2. A. VARIUM

Plants medium-size to ± robust, yellowish to dark green or blackish, in soft, thin or dense, intricate mats, often floating or submerged; stems freely branched; leaves rather distant on stem, often appearing ± complanate, ovate-lanceolate, slenderly acuminate, decurrent; cells oblong-linear to linear, eight to fifteen times as long as wide; costa ending in midleaf or extending into acumen.

Habitat: Streams, ponds, swamps, depressions in low forests;

1. **Amblystegium riparium**
(Hedw.) B.S.G.
Figure 51, D–E

Leptodictyum riparium (Hedw.) Warnst.

L. sipho (P.-Beauv.) Broth.

on wet substrates or floating or submerged; often stranded by receding waters.

Range: Widespread in the Northern Hemisphere.

Gulf Coast Distribution: Florida to Texas.

This moss is widespread and common in our area. Its large leaves make it easy to recognize. It often occurs in huge abundance in favorable situations. I have found it in great floating masses in an old water-filled concrete cellar and an abandoned iron syrup kettle in a forest. The habitat of *A. riparium* is much more toward the aquatic than is that of *A. varium*, even though it thrives on wet substrates on the forest floor just as well as actually in water. Some colonies of *A. riparium* growing in water may have, at first sight, very much the aspect of *Fontinalis*. See comments under *A. varium*.

2. Amblystegium varium
(Hedw.) Lindb.
Figure 51, F–G

Plants small, slender, dull, yellowish green to brownish, in soft or ± rigid, dense or thin mats; leaves crowded to rather loose, ovate-lanceolate, acuminate, entire to slightly serrulate; cells short, oblong to rhombic, mostly about two to three times as long as wide; stem leaves often conspicuously larger than branch leaves; costa ending in acumen, often sinuous above.

> **Habitat:** Soil, humus, stumps, logs, tree bases, rocks, old bricks; low forests; wet areas.
> **Range:** Widespread in the Northern Hemisphere.
> **Gulf Coast Distribution:** Florida to Texas.

The smaller leaves and shorter cells distinguish this drab little moss easily from *A. riparium*. This species does not usually grow in habitats as wet as those of *A. riparium*, but is most frequently found in damp sites.

BRACHYTHECIACEAE Broth.

Plants slender to moderately robust, dark to yellowish green, mostly in ± dense mats, often glossy, mostly firm; stems mostly freely branched, branches often short; leaves slightly to very concave, mostly serrate on margins, sometimes plicate, broadly ovate to ovate-lanceolate, abruptly or gradually acuminate or attenuate, sometimes with shorter, blunter points; cells elongate, apical ones sometimes short and different from median cells, basal cells some-what differentiated, often subquadrate in distinct alar groups; costa strong, single, ending at midleaf or beyond, tip sometimes project-ing as a dorsal tooth. Setae elongated, sometimes papillose; cap-sules nearly erect and symmetric to curved and inclined; peristome double, cilia present or absent; operculum conic to rostrate; ca-lyptra cucullate, mostly naked, sometimes hairy.

The Brachytheciaceae are mostly woodland plants of mesic to damp sites, generally occupying drier habitats than members of the Amblystegiaceae. The capsules of the Brachytheciaceae are shorter and thicker, and less curved, than those of the Amblystegiaceae, and the peristome teeth tend to be reddish brown in the former as opposed to yellowish brown in the latter. Five genera of this family occur on the Gulf Coast.

1. Plants robust; stems and branches distinctly turgid; leaves very concave, abruptly acuminate — 3. BRYOANDERSONIA
1. Plants slender to moderately robust; stems and branches not evidently turgid; leaves not conspicuously concave, variously acute or acuminate but mostly not abruptly so — 2
  2. Leaves acute or obtuse; apical cells of branch leaves short and different from median cells; costa tip distinctly projecting as a tooth on dorsal side of leaf — 5. EURHYNCHIUM
  2. Leaves acuminate; apical cells of branch leaves elongate, not much different from median cells; costa not usually projecting as a dorsal tooth — 3
3. Plants small, delicate, slender, growing on tree trunks; leaves not plicate; seta papillose; calyptra with delicate hairs — 1. HOMALOTHECIELLA
3. Plants slender to moderately robust, not delicate, growing on soil, rock, or tree bases; leaves often plicate; seta mostly smooth; calyptra naked — 4
  4. Leaves usually plicate, not complanate, acute to short-acuminate, apex not twisted; operculum conic or convex-conic — 2. BRACHYTHECIUM
  4. Leaves not plicate, mostly complanate, long-acuminate, with apex twisted; operculum rostrate — 4. RHYNCHOSTEGIUM

1. HOMALOTHECIELLA (Card.) Broth.

Homalotheciella subcapillata (Hedw.) Broth.
Figure 51, H–J

Homalotheciella fabrofolia (Grout) Broth.

Plants small, pale yellowish green, in thin intricate mats; stems creeping; branches short, ± erect, loosely foliated, often curved, close together on stems; leaves about 1 mm long, somewhat concave, rather abruptly acuminate into a long, slender tip, margins serrate above, sometimes all around; median cells elongate; alar cells quadrate, numerous, forming conspicuous areas; costa single, ending about midleaf, sometimes faint. Setae papillose; capsules erect, nearly symmetric; peristome pale; operculum rostrate; calyptra covered with sparse, delicate hairs.

Habitat: Deciduous forests; mostly on tree bark, rarely on tree bases and logs; soft-barked trees are the most common substrate.
Range: Eastern United States.
Gulf Coast Distribution: Florida to eastern Texas.

This delicate little moss is not often collected, doubtless because of its inconspicuous nature. It can be recognized in the field by its habitat and by its fuzzy appearance, which is due to the loosely arranged leaves with their long, slender tips, crowded on the short branches. *Clasmatodon parvulus* is somewhat similar but is darker and has its leaves more appressed and lacking the long, slender tips. Elms, white oaks, magnolias, and sweet gum are frequent substrate trees.

2. BRACHYTHECIUM B.S.G.

Plants slender to fairly robust, in green to yellowish green, usually dense, often glossy mats; stems mostly creeping, densely foliated, with numerous, short, densely foliated branches; stems and branches often mostly pointed ± in one direction giving the mat a "combed" appearance; leaves mostly plicate, serrate on the margins; cells linear; alar cells usually ± differentiated, subquadrate to hexagonal; costa ending below the leaf apex. Infrequently fruiting in our area; setae smooth to papillose; capsules erect to inclined; operculum conic to conic-rostrate.

Brachythecium is a complex assemblage of plants. Quite a few species are attributed to eastern North America, but many of them are quite similar. The following key should separate most of the specimens encountered along the Gulf Coast.

1. Leaves strongly plicate; branch tips curved; upper leaves usually appearing homomallous — 2. B. OXYCLADON
1. Leaves not or only somewhat plicate; branch tips straight; upper leaves not homomallous — 2
   2. Plants pale, silvery; stems and branches slender; leaves appressed; alar cells small, dense, quadrate in conspicuous areas; leaves often slightly plicate — 1. B. ACUMINATUM
   2. Plants yellowish green; stems and branches broad, loosely foliated; alar cells large, lax; leaves not plicate — 3. B. ROTAEANUM

1. Brachythecium acuminatum (Hedw.) Aust.
Figure 52, A–B

Plants glossy, pale to yellowish green; stems and branches slender, pointed, terete, usually all pointing in the same direction; leaves ovate-acuminate to triangular-acuminate, somewhat biplicate or smooth; cells linear, alar ones quadrate, often numerous and forming conspicuous areas. Capsules essentially erect and symmetric; cilia of endostome absent or rudimentary.

Habitat: Mostly on tree bases in deciduous forests.
Range: North America east of the Rockies.
Gulf Coast Distribution: Florida to Texas.

Brachythecium acuminatum is fairly common along the Gulf Coast but is not often collected with sporophytes. Its tree base habitat, "combed" appearance, and pale, silky aspect combine to make it recognizable in the field. It also is found occasionally on soil and rock. *Brachythecium biventrosum* (C.M.) Jaeg. & Sauerb.

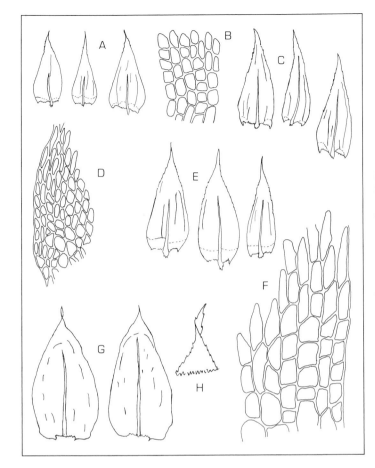

Fig. 52. *A–B*: **Brachythecium acuminatum.** *A*, leaves (× 15). *B*, alar region (× 231). *C–D*: **Brachythecium oxycladon.** *C*, leaves (× 15). *D*, alar region (× 231). *E–F*: **Brachythecium rotaeanum.** *E*, leaves (× 15). *F*, alar region (× 231). *G–H*: **Bryoandersonia illecebra.** *G*, leaves (× 15). *H*, leaf tip (× 55).

and *B. splendens* Aust. have been attributed to the Gulf Coast. However, it is not clear that either differs appreciably from the concept of *B. acuminatum*; both names have been placed in the synonymy of *B. acuminatum* by Crum & Anderson (1981).

Plants slender to rather robust, in dense or loose mats, green to yellowish green; stems usually elongate and lax, often not much branched; branches short; both stems and branches commonly curved at the tips and with the upper leaves ± secund; leaves triangular-acuminate or lanceolate-acuminate, usually distinctly plicate; cells linear, alar ones usually small, dense, subquadrate, not forming conspicuous areas. Capsules inclined, curved, endostome with cilia.

Habitat: Mostly on soil in forests; ravines, stream banks, gullies, occasionally on rotted logs and rocks.
Range: Eastern North America.
Gulf Coast Distribution: Florida to Texas.

2. Brachythecium oxycladon
(Brid.) Jaeg. & Sauerb.
Figure 52, C–D

This species is usually easy to recognize, even in the field, by its plicate leaves, usually ± secund at the branch and stem tips, by the often curved branch and stem tips, and by the soil habitat. It often forms large, thick mats in favorable habitats, especially on stream banks. It is usually easily distinguished from *B. rotaeanum* by the nonplicate leaves and large, clear, thin-walled alar cells of the latter.

Two similar species—*B. plumosum* (Hedw.) B.S.G. and *B. rutabulum* (Hedw.) B.S.G.—are attributed to the Gulf Coast. Both have the seta papillose, at least in the upper part, whereas the seta is smooth in *B. oxycladon* and *B. rotaeanum*. I have not seen any convincing material of either *B. plumosum* or *B. rutabulum* from the Gulf Coast.

3. Brachythecium rotaeanum
De Not.
Figure 52, E–F

Similar in some respects to *B. oxycladon* but differing, among other ways, in the more loosely foliated stems and branches, larger, clear, thin-walled alar cells, nonplicate leaves, and its habitat on rotted logs and tree bases rather than soil and rock.

> **Habitat:** Rotted logs, tree bases; wet forests.
> **Range:** Southeastern United States; Europe.
> **Gulf Coast Distribution:** Florida to eastern Texas.

This species is not at all common on the Gulf Coast. However, it may be passed over by collectors because it is not a conspicuous moss and occupies rather obscure habitats. I have seen only a few collections of it from our area. See comments under *B. oxycladon*.

BRACHYTHECIUM SALEBROSUM (Web. & Mohr) B.S.G. has been attributed to the Gulf Coast; however, I have not seen any convincing specimens from our area. It is not included in our range by Crum and Anderson (1981).

3. BRYOANDERSONIA Robins.

Bryoandersonia illecebra
(Hedw.) Robins.
Figure 52, G–H

Cirriphyllum illecebrum
(Hedw.) L. Koch

Plants robust, green to yellowish green, in soft, deep mats; stems and branches turgid, densely foliated; leaves very broad, deeply concave, about 2 mm long, abruptly apiculate into a short, often twisted tip, serrate on the margins; cells elongate, somewhat porose, smaller and denser at basal angles; costa extending beyond midleaf, sometimes faint or nearly lacking. Infrequently fruiting; capsules curved and inclined; operculum long-rostrate.

> **Habitat:** Banks and ledges in forests; sometimes on tree bases.
> **Range:** Eastern North America.
> **Gulf Coast Distribution:** Florida to eastern Texas.

This handsome moss is one of our commonest woodland species; it is especially abundant on ravine banks and along streams in deciduous forests. It is very easy to recognize because of its large size and turgid branches bearing deeply concave leaves with abruptly pointed tips.

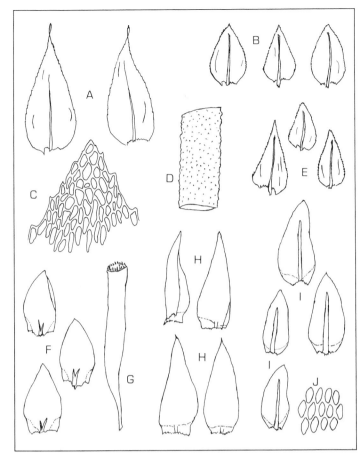

Fig. 53. *A,* **Rhynchostegium serrulatum,** leaves (× 15). *B–D:* **Eurhynchium hians.** *B,* leaves (× 15). *C,* leaf tip (× 231). *D,* portion of seta (× 55). *E,* **Eurhynchium pulchellum,** leaves (× 15). *F–G:* **Entodon seductrix.** *F,* leaves (× 15). *G,* capsule (× 15). *H,* **Entodon macropodus,** leaves (× 15). *I–J:* **Stereophyllum radiculosum.** *I,* leaves (× 15). *J,* median cells (× 231).

Plants medium-size to robust, glossy, green to yellowish green, in thin or dense mats, rather soft; stems and branches usually flattened, sometimes almost julaceus, rather loosely foliated; leaves about 2 mm long, ovate-lanceolate, apex slenderly acuminate, usually twisted, margins serrate; cells elongate, basal ones smaller; costa ending above midleaf, sometimes terminating in a small dorsal tooth. Capsules inclined and strongly curved; operculum strongly rostrate.

Habitat: Soil, rotted wood, logs, tree bases, mostly in wet forests.

Range: Eastern North America; American tropics.

Gulf Coast Distribution: Florida to Texas.

This moss is fairly common throughout much of our area, especially in the eastern part, and is usually easy to recognize. Small forms may be mistaken at first glance for *Isopterygium tenerum,* but that species lacks a costa in its leaves. (The costa of *Rhynchostegium* can be easily seen in the field with a hand lens.) The

4. RHYNCHOSTEGIUM B.S.G.

Rhynchostegium serrulatum (Hedw.) Jaeg. & Sauerb. Figure 53, A stems and branches of *R. serrulatum* are not always complanate, but even so there is no other moss in our area with such long-acuminate leaves having the apex twisted. *Rhynchostegium serrulatum* usually grows in rather wet sites but may also be found on tree bases or soil in fairly dry situations.

5. EURHYNCHIUM B.S.G.   Moderately robust plants, dark to yellowish green, glossy or dull, usually growing in mats; stems freely branched; stem leaves not much different from branch leaves; leaves broadly ovate to lanceolate, serrate, sometimes complanate, apical cells short and quite different from the elongate median cells; costa reaching midleaf or beyond, terminating in a distinct tooth on the dorsal surface of the leaf. Infrequently fruiting in our area; setae smooth or papillose; capsules inclined, asymmetric; operculum long-rostrate.

Two species of *Eurhynchium* occur in our area; one is rather weedy and common. The long rostrum of the operculum, short terminal cells of the leaves, and the dorsal tooth formed at the tip of the costa help identify the genus.

Seta papillose; branches loosely foliated, attenuate at the tips;
 branch leaves broadly ovate, acute   1. E. HIANS
Seta smooth; branches densely foliated, blunt at the tips; branch
 leaves ovate-lanceolate, broadly acute to obtuse   2. E. PULCHELLUM

---

1. Eurhynchium hians (Hedw.) Sande-Lac.
Figure 53, B–D

Oxyrrhynchium hians (Hedw.) Jenn.

Plants slender to fairly robust, in loose, soft mats, dark to yellowish green; stems freely branched; branches loosely foliated, usually attenuate at the tips; leaves serrate, frequently complanate, broadly ovate, apex acute; apical cells short, very different from elongate median cells; costa with a distinct tooth dorsally at its tip. Setae distinctly papillose.

> **Habitat:** Soil and rock in ravines and wet forests; moist sites in general.
> **Range:** Eastern North America; Europe; Asia.
> **Gulf Coast Distribution:** Florida to Texas.

Eurhynchium hians is one of our commonest mosses. It is frequently found in shady places in lawns and around house foundations. As a general rule it occupies damper sites than does *E. pulchellum*. Its papillose seta will distinguish it easily from *E. pulchellum*; if the specimen is sterile, its darker color, looser habit, attenuate branch tips, and broad leaves with acute apex will identify it. The differentiated cells of the leaf apex and the costa projecting as a dorsal tooth will distinguish it from other members of the Brachytheciaceae. I have found this moss repeatedly over the years as a component of the nest of the Carolina wren.

Plants moderately robust, rather dull, yellowish green, in usually dense, springy mats; stems much-branched, the branches mostly short and ascending, often mostly pointed in the same direction, densely foliated, tips rather blunt; leaves serrate, obtuse or broadly acute at apex, apical cells quite different from the elongate median cells; costa ending in a dorsal tooth. Setae smooth.

> **Habitat:** Soil and rocks, tree bases, rotting stumps; banks and ravines in deciduous forests.
> **Range:** Circumpolar; widespread in North America; Mexico; Guatemala; northern South America.
> **Gulf Coast Distribution:** Alabama to Texas. Not yet reported from Florida but likely to be found in the Panhandle area.

This moss is more common and abundant in upland areas of the Gulf Coast than in the lower, coastal parts, and it occupies drier sites in general than *E. hians*. See comments under the latter species.

2. **Eurhynchium pulchellum**
(Hedw.) Jenn.
Figure 53, E

ENTODONTACEAE Kindb.

Plants glossy, in thin or dense, spreading mats; stems creeping to ascending, irregularly to pinnately branched, stems and branches julaceus or, sometimes, flattened; leaves broadly to narrowly ovate-lanceolate, acute, mostly entire except at apex; cells mostly linear, smooth, quadrate in alar regions in usually distinct groups; costa single, short and double, or lacking. Setae elongate; capsules mostly erect and symmetric; peristome mostly double; operculum conic to rostrate; calyptra cucullate, naked.

   This family is represented along the Gulf Coast by only one genus, *Entodon*, with two species.

Our two species of *Entodon* are forest mosses having nearly terete to strongly flattened stems, conspicuously quadrate alar cells, the costa lacking or short and double, and a double peristome. They are all glossy plants of medium to large size. One species, *E. seductrix*, is very common and abundant over much of our area; the other species, *E. macropodus*, is less common.

ENTODON C.M.

Stems and branches narrow, mostly julaceus but sometimes ± flattened; seta red to brownish red	1. E. SEDUCTRIX
Stems and branches broad, strongly flattened; seta yellow	2. E. MACROPODUS

1. Entodon seductrix (Hedw.) C.M.

Figure 53, F–G

Plants green to yellowish green, in thin or thick, ± glossy mats; stems terete or often somewhat flattened, creeping or ascending, arching and rooting at tips, irregularly to subpinnately branching; leaves deeply concave, broadly ovate, acute to apiculate, margins entire, sometimes serrulate at apex; cells fusiform to linear, quadrate in conspicuous areas in alar regions; costa double, short or reaching ⅓ leaf length. Sporophytes commonly produced in abundance; setae dark red to red-brown; capsules cylindric, slightly asymmetric-curved; operculum obliquely rostrate.

> **Habitat:** Tree bases, logs, stumps, soil, rock; mesic to wet forests.
> **Range:** Eastern North America.
> **Gulf Coast Distribution:** Florida to Texas.

This is a weedy little moss that is very common in most of our area. Although its most frequent habitat is tree bases, it is commonly found on soil and logs along the Gulf Coast. The terete or only somewhat flattened stems and branches distinguish it from our other species of *Entodon*. This moss sometimes grows in great, often glossy sheets on logs, tree bases, and rocks. It seems to be tolerant of pollution and is a city moss in the sense that it is often abundant in cities in parks and on tree bases along streets.

2. Entodon macropodus (Hedw.) C.M.

Figure 53, H

Plants yellowish to yellowish green, glossy, in thin, spreading mats; stems broad, strongly flattened, irregularly branched; leaves oblong-ovate, acute, strongly complanate, often folded lengthwise, somewhat asymmetric, not concave, entire except at apex; cells linear to long-linear, quadrate to somewhat inflated in distinct groups at basal angles, the two groups often unequal in size; costa very faint, short and double or lacking. Sporophytes commonly produced; setae pale yellow; capsules cylindric, slightly asymmetric; operculum obliquely conic-apiculate to stoutly conic-rostrate.

> **Habitat:** Tree bases, logs, roots, rocks; mesic to wet forests.
> **Range:** Southern United States; tropical America; Asia.
> **Gulf Coast Distribution:** Florida to eastern Texas.

This handsome moss is rather common in rich, mesic forests over much of our area. The glossy, highly flattened stems with yellow setae are distinctive. It sometimes forms large, very conspicuous mats.

ENTODON CLADORRHIZANS (Hedw.) C.M. has been reported at various times from along the Gulf Coast, but the reports seem to be based on misidentifications. This species apparently does not occur in our area.

PLAGIOTHECIACEAE (Broth.) Fleisch.

Plants small to somewhat robust, often glossy, in thin or dense, spreading mats; stems mostly creeping and irregularly branched, ± flattened: leaves ovate to elliptic or lanceolate, acute to acuminate, mostly complanate; cells mostly linear, or rhombic to oblong-linear, smooth or sometimes papillose, sometimes differentiated at basal angles; costa short and double, single, or lacking. Setae elongate; capsules erect to horizontal, mostly ± asymmetric, peristome double; operculum convex-conic to rostrate; calyptra cucullate, naked.

This family is represented along the Gulf Coast by only one genus, with a single species in our area. Breen (1963) reported *Plagiothecium sylvaticum* (Brid.) B.S.G. [= *P. caviifolium* (Brid.) Iwats.] from the Florida Panhandle, but Ireland (1969) believes that the specimens upon which Breen based her report actually represent a species in another genus, *Taxiphyllum alternans* (Card.) Iwats.

STEREOPHYLLUM Mitt.

Stereophyllum radiculosum (Hook.) Mitt.
Figure 53, I–J

Plants slightly glossy, yellowish to brownish green, in thin, sparse, spreading mats; stems very irregularly branched, julaceus or ± flattened, creeping; lateral leaves smaller than dorsal leaves; leaves ± complanate, broadly ovate to oblong-ovate, broadly acute, erect-appressed when dry, spreading when moist, somewhat asymmetric; cells short, rhombic, smooth to distinctly papillose, quadrate in large, conspicuous areas at basal angles; costa single, very stout, extending nearly to base of acumen. Capsules erect to inclined, slightly asymmetric; operculum obliquely conic-rostrate.

> **Habitat:** Tree bases, stumps, roots, limestone; dry, ± open, low forests.
> **Range:** Southern United States; tropical America.
> **Gulf Coast Distribution:** Southern Texas.

This moss is fairly common, but rarely abundant, in central and southern Texas. It also occurs in the United States in southern Florida. Its most frequent habitat in Texas is on the bases of live oak trees.

SEMATOPHYLLACEAE Broth.

Plants slender to somewhat robust, mostly glossy, green to yellowish green or golden, in mostly soft, spreading mats; stems mostly creeping and irregularly branched; leaves crowded, often ± complanate, often homomallous or secund and falcate, mostly ovate to lanceolate; cells mostly oblong-linear to linear but sometimes shorter, smooth to papillose, mostly colored at the leaf base and distinctly differentiated (usually conspicuously inflated) at the basal angles; costa short and double or lacking. Setae elongate; capsules small, mostly inclined to pendulous and ± asymmetric, sometimes erect and symmetric; peristome double; operculum mostly rostrate; calyptra naked, mostly cucullate.

The Sematophyllaceae are a large family with many representatives in the tropics. The family is represented along the Gulf Coast by *Sematophyllum*, with three species, and *Brotherella*, with one species. *Sematophyllum adnatum*, growing on trees and logs, is common throughout most of our area. The other species are infrequent or rare on the Gulf Coast.

Leaves falcate-secund, serrate above	1. Brotherella
Leaves straight, entire	2. Sematophyllum

1. BROTHERELLA Loeske ex Fleisch.

Brotherella recurvans (Michx.) Fleisch.
Figure 54, A–B

Plants very glossy, green to golden green or yellowish, in dense, soft mats; stems subpinnately branched, the branches prostrate and flattened, not curved upward at the tips; leaves falcate-secund, or only the tips falcate and secund, oblong-ovate, slenderly acuminate; margins usually sharply serrate above; cells linear above, shorter and yellow at the base, alar cells inflated in conspicuous groups of about 4–8 cells; costa lacking. Capsules nearly erect to inclined, asymmetric; operculum obliquely rostrate.

> **Habitat:** Soil, rock, humus, tree bases, rotted wood and logs; mostly in wet forests.
> **Range:** Widespread in eastern North America but mostly in the northeast; Japan.
> **Gulf Coast Distribution:** Eastern Texas.

This species was reported long ago from Louisiana, but I have not seen the specimen. Stoneburner and Wyatt (1979) reported it from Liberty County, Texas, where it occurred on a decaying log in a wet hardwood forest. It is evidently very rare along the Gulf Coast. The complanate, falcate-secund leaves distinguish it from *Sematophyllum*, and its golden-glossy appearance, complanate leaves, and yellow basal cells distinguish it from *Hypnum*.

2. SEMATOPHYLLUM Mitt.

Plants slender, soft, mostly glossy, creeping and forming mats; stems ± irregularly branched with the branches often short and curved upward at the tips; leaves ovate to lanceolate, acute to acuminate, entire, often concave; alar cells conspicuously inflated in distinct groups; upper and median cells ± rhombic to fusiform or linear; basal cells mostly yellow or golden; margins entire to serrate; costa short and double or lacking. Capsules erect and symmetric to inclined or pendulous and asymmetric.

Plants of this genus are usually easy to recognize due to the

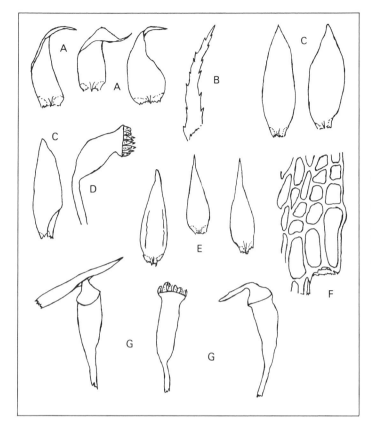

Fig. 54. *A–B*: **Brotherella recurvans.** *A*, leaves (× 21). *B*, leaf tip (× 86). *C–D*: **Sematophyllum demissum.** *C*, leaves (× 21). *D*, capsule (× 21). *E–G*: **Sematophyllum adnatum.** *E*, leaves (× 21). *F*, alar region (× 328). *G*, capsules, one with unshed calyptra (× 14).

inflated and often colored alar cells. *Brotherella recurvans* also has inflated alar cells, but the leaves are complanate and falcate-secund. Some species of *Hypnum* also have inflated alar cells, but in that genus the leaves are mostly falcate-secund.

1. Capsule erect; plants mostly growing on bark and logs; common	2. S. ADNATUM
1. Capsule distinctly inclined to horizontal or pendulous; plants growing on rocks, roots, or tree bases; infrequent	2
2. Plants growing on rocks; leaf cells elongate, narrow, six to eight times as long as wide.	1. S. DEMISSUM
2. Plants growing on roots and tree bases; leaf cells short, broad, two to six times as long as wide.	3. S. CAESPITOSUM

Plants medium-size, glossy, yellowish brown to brownish green, in spreading, sordid mats; stems irregularly branched, creeping or ascending; leaves lanceolate, concave, acute to short-acuminate; cells elongate, narrow, more than six times as long as wide, walls often thickened and ± pitted, cells along the insertion colored, inflated in alar areas. Capsules inclined to horizontal or pendulous, ± asymmetric.

1. Sematophyllum demissum (Wils.) Mitt.
Figure 54, C–D

Sematophyllum carolinianum (C.M.) E. G. Britt.

Habitat: Sandstone; mostly along streams in forests.
Range: Eastern North America.
Gulf Coast Distribution: Florida to eastern Texas.

Although this species has not yet been reported from southern Alabama or the Florida Panhandle, it probably occurs there. It seems to be restricted in our area to sandstone, mostly along streams, where it is often abundant. The habitat and inclined capsules distinguish it easily from *S. adnatum*, which also may grow on rock, and its long, narrow cells (and habitat) distinguish it from *S. caespitosum*.

2. Sematophyllum adnatum
(Michx.) E. G. Britt.
Figure 54, E–G

Plants small, glossy, green to yellowish green, in soft, dense or thin mats; stems freely branched, branches usually short and turned up at tips; leaves lanceolate, acuminate; cells elongate, narrow, eight to twelve times as long as wide, conspicuously inflated in alar areas. Capsules erect and symmetric.

Habitat: Trees, stumps, logs, sometimes on rocks; mesic to dry forests.
Range: Eastern United States; Mexico; Central and South America.
Gulf Coast Distribution: Florida to Texas.

This might be the most common moss along the Gulf Coast; only *Clasmatodon parvulus*, also a bark-inhabiting moss, could be a contender for first place. *Sematophyllum adnatum* is easy to recognize since it produces sporophytes freely and the erect capsules are distinctive. *Clasmatodon parvulus* also has erect capsules, but the plants are smaller and the leaves lack inflated alar cells and possess a well-developed costa, among other differences.

3. Sematophyllum caespitosum (Hedw.) Mitt.
Figure 55, A–C

Plants in thin, dark to yellowish green mats; stems creeping to ± ascending, rather loosely foliated; leaves ovate to ovate-lanceolate, acute, short-pointed; cells short, rather broad, rhombic to short-fusiform, two to six times as long as wide, often colored across insertion, conspicuously inflated in alar regions. Capsules inclined to horizontal.

Habitat: Tree bases and exposed roots; low forests.
Range: Southern United States; tropical America.
Gulf Coast Distribution: Louisiana and Mississippi.

This inconspicuous little moss is rare in our area and apparently is known along our part of the Gulf Coast only from low, humid forests in southeastern Louisiana and southern Mississippi. It doubtless will be found in southern Alabama and the Florida Panhandle; it is fairly common in southern Florida. Its most frequent habitat in Louisiana is on exposed roots of beech trees along streams and in wet forests. See comments under *S. demissum*.

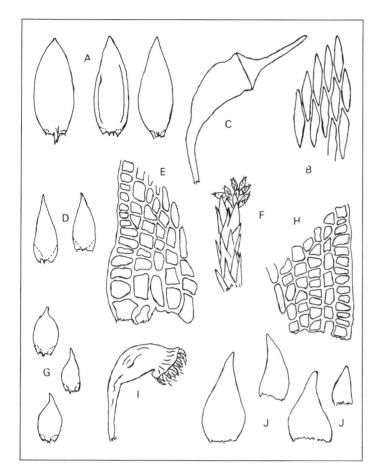

Fig. 55. *A–C:* **Sematophyllum caespitosum.** *A,* leaves (× 21). *B,* median cells (× 328). *C,* capsule (× 14). *D–F:* **Platygyrium repens.** *D,* leaves (× 21). *E,* alar region (× 328). *F,* branch tip with brood bodies (× 21). *G–I:* **Homomallium adnatum.** *G,* leaves (× 21). *H,* alar region (× 328). *I,* capsule (× 21). *J,* **Platydictya confervoides,** leaves (× 86).

HYPNACEAE Schimp.

Plants very small to robust, often glossy, growing in mats; stems mostly creeping, irregularly to pinnately branched; paraphyllia mostly lacking; branch and stem leaves mostly similar, often secund to falcate-secund, sometimes complanate, mostly acuminate; cells mostly linear (sometimes short), smooth, or rarely papillose by projecting cell ends or by papillae over the lumens; cells at basal angles mostly differentiated, quadrate to inflated; costa short and double or lacking. Sporophytes lateral; setae elongate; capsules inclined to horizontal, mostly curved, sometimes erect and symmetric; peristome mostly double; operculum convex to rostrate; calyptra cucullate, mostly naked.

The Hypnaceae are a large family of mostly glossy, terrestrial, mat-forming mosses. Some are corticolous, others grow on stumps and logs or rocks. Representatives of seven genera are known from our area.

1. Leaf cells papillose dorsally; stem leaves ± differentiated from branch leaves — 7. CTENIDIUM
1. Leaf cells smooth; stem and branch leaves not particularly different from one another — 2
    2. Plants very small and slender; leaves 0.5 mm long or shorter; habitat on limestone — 3. PLATYDICTYA
    2. Plants small to robust; leaves longer than 0.5 mm; habitat various — 3
3. Leaves strongly falcate-secund; branching regularly pinnate or nearly so — 4. HYPNUM in part
3. Leaves not or only slightly secund and falcate; branching mostly irregular — 4
    4. Alar cells inflated in conspicuous groups — 4. HYPNUM in part
    4. Alar cells not conspicuously inflated — 5
5. Plants flattened, leaves complanate — 6
5. Plants not flattened, leaves not complanate — 7
    6. Apical cells of leaves conspicuously shorter than median cells; pseudoparaphyllia leaflike — 6. TAXIPHYLLUM
    6. Apical cells of leaves not conspicuously different from median cells; pseudoparaphyllia mostly filamentous — 5. ISOPTERYGIUM
7. At least some branches bearing clusters of leafy brood bodies in axils of upper leaves; leaf margins recurved below; cells elongate — 1. PLATYGYRIUM
7. Brood bodies lacking; leaf margins erect; cells short — 2. HOMOMALLIUM

1. PLATYGYRIUM B.S.G.

Platygyrium repens (Brid.) B.S.G.
Figure 55, D–F

Plants small, glossy, dark green to yellowish or brownish, in tight low mats; stems creeping, freely and usually closely branched, branches short, often curved, erect-ascending, at least some usually bearing conspicuous clusters of small, leafy brood bodies in axils of upper leaves; leaves erect to appressed when dry, concave, ovate-lanceolate, about 1 mm long, abruptly short-acuminate, margins reflexed in lower ⅔–½; cells oblong-rhombic to fusiform or linear, much shorter in acumination, quadrate alar cells conspicuous; costa short and double. Capsules erect and symmetric; peristome double; operculum conic-rostrate; not or only rarely producing sporophytes in our area.

Habitat: Logs, stumps; also on tree trunks, bases and branches; in mesic to rather dry upland forests.
Range: Widespread in the Northern Hemisphere; eastern North America.
Gulf Coast Distribution: Florida to eastern Texas.

Howard Crum (1976) aptly described the colonies of this moss as "blackish-green mats with a curious oily sheen." Once one comes to know *P. repens*, its colonies can be recognized at some distance just by the color and habitat. The plants are sometimes more yellowish than blackish. The leafy brood bodies are almost always conspicuously present at the branch tips and can be seen readily with a hand lens. *Platygyrium* differs from *Homomallium* in its erect capsules, reflexed leaf margins, longer cells, and brood bodies. In our area, *Platygyrium* characteristically grows on logs.

PLATYGYRIUM FUSCOLUTEUM Card. was reported from our area in Texas by Whitehouse and McAllister (1954); however, I have seen the specimen upon which the report is based, and it is *P. repens*. This species could possibly be expected in southern Texas.

2. HOMOMALLIUM (Schimp.) Loeske

Homomallium adnatum
(Hedw.) Broth.
Figure 55, G–I

Plants slender, dark green, in thin spreading mats; stems creeping, irregularly branched; branches short, straight; leaves concave, ovate-lanceolate, short-acuminate, erect when dry, rarely slightly secund at branch tips, 0.5–0.8 mm long, margins erect, entire or a little serrulate near apex; cells rhombic, smooth, quadrate in conspicuous areas at basal angles. Capsules yellowish orange, usually strongly curved when dry, 1–1.5 mm long; operculum stoutly apiculate.

> **Habitat:** Rocks, sometimes tree bases; mesic to dry upland forests.
> **Range:** Eastern North America.
> **Gulf Coast Distribution:** Texas.

Homomallium adnatum is a rather nondescript little moss when seen in the field. It resembles *Clasmatodon parvulus* macroscopically, but that species has erect capsules and costate leaves. It also is generally similar to *Platygyrium* but differs in lacking brood bodies, in having erect leaf margins, and in its shorter leaf cells. I have only seen a single authentic specimen from our area, from Nacogdoches County, Texas; however, this species could well occur across our area at inland sites from Texas eastward.

HOMOMALLIUM MEXICANUM Card. has been reported from our area in Texas, but I have not seen any authentic material and doubt that it occurs in eastern Texas. It differs from *H. adnatum* in having more slenderly acuminate leaves that are usually decidedly homomallous and in its less curved capsule.

3. PLATYDICTYA Berk.

Platydictya confervoides
(Brid.) Crum
Figure 55, J

Amblystegiella confervoides
(Brid.) Loeske

Plants very small, slender, dull, brownish green to darker, in thin mats, rather rigid; stems prostrate or ascending, irregularly branched; leaves very small, erect-spreading when moist, lanceolate to ovate-lanceolate, acuminate, 0.15–0.3 mm long, entire or slightly serrulate; cells oblong to rhombic, smooth, subquadrate at basal angles. Capsules inclined to nearly erect; operculum stoutly apiculate.

> **Habitat:** Creeping on limestone.
> **Range:** North America; Europe.
> **Gulf Coast Distribution:** Jackson County, Florida.

This tiny and inconspicuous moss is known in our area only from the Florida Caverns locale near Marianna, Florida, but could well occur on limestone elsewhere in our area. It is distinctive in its small size.

PLATYDICTYA SUBTILE (Hedw.) Crum has been reported from our area in Baldwin County, Alabama, but I have not been able to see the specimen and doubt that it occurs so far south. It has slightly larger leaves than *P. confervoides* and grows on tree bark, and its capsules are essentially erect rather than inclined.

4. HYPNUM Hedw.

Plants slender to moderately robust, firm, in soft, intricate mats, green to yellowish or golden green, glossy; stems mostly creeping, irregularly to pinnately branched, stems and branches mostly hooked downwards at tips; leaves crowded, often apparently in two rows, sometimes complanate, ± secund to strongly falcate-secund, ± concave, ovate- to triangular-lanceolate, acute to long-acuminate; margins entire to serrulate; median cells mostly linear, smooth, short at apex, quadrate and with thickened, porose walls to inflated at basal angles; costa lacking or short and double. Sporophytes rarely produced in our area; capsules inclined to horizontal or nearly erect, mostly curved and symmetric; peristome double; operculum conic to apiculate or somewhat rostrate; calyptra naked.

Three species of *Hypnum* occur along the Gulf Coast; two of them are rather common and widespread. Our species of *Hypnum* can be recognized by the pinnate (or nearly so) branching and the mostly falcate-secund leaves.

1. Stems mostly ± flattened, irregularly branched; leaves complanate, neither strongly falcate nor secund, apex broad and short; alar cells inflated in conspicuous areas .... 1. H. LINDBERGII
1. Stems not evidently flattened, mostly pinnately branched; leaves not complanate, strongly falcate-secund, apex acuminate; alar cells not inflated, or only a few inflated and not forming conspicuous areas 2
   2. Pseudoparaphyllia entire; capsule strongly curved; incrassate alar cells few and inconspicuous; 1–3 alar cells inflated .... 2. H. CURVIFOLIUM
   2. Pseudoparaphyllia ± ciliate on margins; capsule nearly erect; incrassate alar cells in conspicuous hyaline to brownish orange groups; inflated alar cells lacking .... 3. H. IMPONENS

1. **Hypnum lindbergii** Mitt.
Figure 56, A–C

Plants glossy, moderate-size, in dense, yellowish to pale green mats; stems creeping or ± ascending, irregularly branched, mostly somewhat flattened; leaves oblong-lanceolate, acute to broadly acuminate, mostly somewhat falcate-secund but not strongly so; margins mostly entire; cells linear, abruptly inflated and thin-walled in conspicuous areas in alar regions. Capsules inclined and curved.

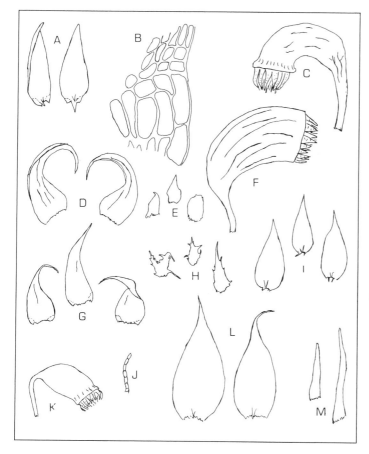

Fig. 56. *A–C:* **Hypnum lindbergii.** *A,* leaves (× 15). *B,* alar region (× 231). *C,* capsule (× 15). *D–F:* **Hypnum curvifolium.** *D,* leaves (× 15). *E,* pseudoparaphyllia (× 55). *F,* capsule (× 15). *G–H:* **Hypnum imponens.** *G,* leaves (× 15). *H,* pseudoparaphyllia (× 55). *I–K:* **Isopterygium tenerum.** *I,* leaves (× 15). *J,* pseudoparaphyllium (× 55). *K,* capsule (× 15). *L–M:* **Isopterygium cuspidifolium.** *L,* leaves (× 15). *M,* pseudoparaphyllia (× 55).

Habitat: Soil, logs, humus, rock; swamps, wet forests, springy places, stream banks.

Range: Widespread in the Northern Hemisphere.

Gulf Coast Distribution: Florida to Texas.

This is our most common species of *Hypnum*. It occupies generally wetter sites than our other species. The broad, scarcely acuminate leaves, only slightly falcate-secund, and conspicuous areas of inflated alar cells make it easy to recognize.

Plants glossy, moderate-size, in dense, yellowish to golden green mats; stems creeping, irregularly to pinnately branched, not flattened; pseudoparaphyllia entire; leaves strongly falcate-secund, long-acuminate, triangular-ovate, margins ± serrulate above; cells linear, a few quadrate and incrassate in inconspicuous groups in alar regions, below these often 1–3 conspicuously inflated, thin-walled hyaline cells. Capsules inclined, strongly curved.

Habitat: Soil, tree bases, stream and ravine banks; mesic to wet forests.

2. **Hypnum curvifolium** Hedw.
Figure 56, D–F

Range: Eastern North America.
Gulf Coast Distribution: Alabama to eastern Texas.

This species is fairly common in much of our area. Although it has not been recorded from the Florida Panhandle, it probably occurs there. It is a species inhabiting upland forests along the Gulf Coast. Its strongly falcate and secund leaves with slender acumination and few inflated alar cells distinguish it easily from *H. lindbergii.* Inflated alar cells are not present on every leaf. In *H. imponens* there are no inflated alar cells, but instead a conspicuous area of thick-walled, hyaline to brownish quadrate cells. When sporophytes are present, the strongly curved capsules of *H. curvifolium* will distinguish it from *H. imponens,* which has a nearly erect capsule. See other comments under *H. imponens.*

3. **Hypnum imponens** Hedw.
Figure 56, G–H

Similar in many respects to *H. curvifolium;* differing most obviously in the more slenderly pointed leaves, lack of inflated alar cells, presence of conspicuous areas of incrassate, hyaline or brownish orange quadrate alar cells, and the nearly erect capsules. This species also has rather abundant multiform, ± ciliate pseudoparaphyllia as opposed to the small, entire pseudoparaphyllia of *H. curvifolium.*

> **Habitat:** Tree bases, logs, humus, soil; mesic forests.
> **Range:** North America; Europe; Asia.
> **Gulf Coast Distribution:** Florida.

I have seen only one specimen of this species from the Gulf Coast; it was collected in Escambia County, Florida, by David Breil. It certainly could be expected elsewhere in our range. See comments under *H. curvifolium.*

5. **ISOPTERYGIUM** Mitt.

Plants small to medium-size, pale to yellowish green or darker, soft, glossy, in thin mats; stems creeping or somewhat ascending, irregularly branched, ± flattened; pseudoparaphyllia mostly filamentous; leaves lanceolate to broadly ovate-lanceolate, often ± complanate, sometimes ± secund; margins entire to serrulate; cells smooth, linear, a few subquadrate to quadrate and sometimes lax or with thickened walls at basal angles; costa short and double or lacking. Capsules small, nearly erect to inclined, ± asymmetric; peristome double; operculum conic to apiculate-rostrate.

Two species of *Isopterygium* occur along the Gulf Coast; one, *I. tenerum,* is exceedingly common over most of our area.

Plants small; leaves mostly 1–1.5 mm long and less than 0.5 mm wide; differentiated alar cells few, subquadrate to lax; pseudoparaphyllia mostly filamentous .... 1. I. TENERUM

Plants medium-size; leaves mostly 2–3 mm long, 1 mm wide; alar cells quadrate in several rows; pseudoparaphyllia foliose .... 2. I. CUSPIDIFOLIUM

Plants small, slender, glossy, yellowish green, in thin mats; stems mostly clearly flattened; pseudoparaphyllia filamentous (rarely two cells wide); leaves spreading, sometimes ± secund, ovate-lanceolate, acuminate, mostly 1–1.5 mm long, 0.5 mm wide; margins entire or serrulate above; cells linear, differentiated across insertion, a few in basal angles subquadrate to oblong, lax or with ± thickened walls. Sporophytes commonly produced; capsules inclined to nearly erect, small, asymmetric.

Habitat: Tree bases, stumps, logs, soil, humus; mesic to wet forests and swamps.
Range: Eastern United States; tropical America; Asia.
Gulf Coast Distribution: Florida to eastern Texas.

This little moss is very common throughout much of our area, and is also quite variable in size and appearance. The best field clues to its recognition are the small size, glossy aspect, ± flattened stems, and small, asymmetric and inclined capsules. Species of *Sematophyllum* are somewhat similar in general aspect and may grow in similar habitats, but their leaves always have shorter median cells and conspicuously inflated alar cells. Our common *Sematophyllum adnatum* has erect capsules.

Plants dark to yellowish green, in soft, glossy mats; stems somewhat flattened or ± turgid; pseudoparaphyllia foliose; leaves ascending to ± spreading, ovate- to broadly ovate-lanceolate, rather abruptly slenderly short-acuminate, mostly 2–3 mm long, 1 mm wide; margins serrulate above; cells linear, quadrate in several rows at basal angles; costa lacking to short and double, one branch sometimes elongated to ½–⅓ leaf length. Not producing sporophytes in our area.

Habitat: Limestone in mesic forests.
Range: Southeastern United States; Asia.
Gulf Coast Distribution: Florida.

This rare moss is known in our area only from Jackson and Walton counties, Florida. It also occurs elsewhere in Florida and is known from a few sites in northern Alabama and Tennessee. The size and somewhat turgid stems of the plants cause it to resemble a *Plagiothecium*, but plants of that genus lack pseudoparaphyllia. The differences cited in the key distinguish *I. cuspidifolium* easily from the common *I. tenerum*.

Plants small to rather robust, glossy, yellowish green to green, in ± flat, spreading mats; stems prostrate, not much branched, sometimes mostly all pointed in the same direction; pseudoparaphyllia present and leaflike; leaves crowded on stems or ± distant, complanate, often appearing to be distichous, ovate to ovate-lanceolate, acute to acuminate, concave, often ± asymmetric, mar-

1. Isopterygium tenerum (Sw.) Mitt.
Figure 56, I–K

Isopterygium micans (Sw.) Kindb.

2. Isopterygium cuspidifolium Card.
Figure 56, L–M

Taxiphyllum mariannae (Grout) Breen

6. TAXIPHYLLUM Fleisch.

gins sometimes recurved below, serrulate; median cells linear, apical cells conspicuously shorter, quadrate alar cells few; costa double or lacking, sometimes conspicuous and with one branch reaching ¼ or more leaf length. Not producing sporophytes in our area. Capsules inclined, symmetric or curved; peristomes double; opercula obliquely rostrate.

Two species of *Taxiphyllum* occur along the Gulf Coast. The complanate leaves with the cells in the leaf tip very short relative to the median cells make the genus rather easy to recognize. One species, *T. taxirameum*, is common; the other species is quite rare.

Leaves 1 mm or more wide; plants ± robust, in soft, dense mats in swamps   1. T. ALTERNANS

Leaves less than 1 mm wide; plants slender, in thin, mostly flat mats in drier sites   2. T. TAXIRAMEUM

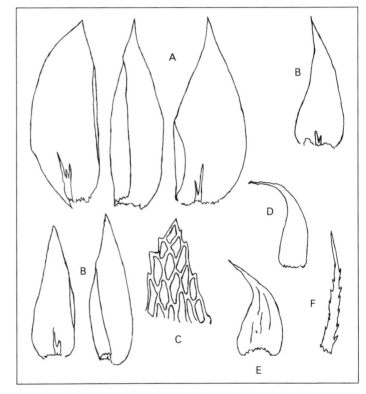

Fig. 57. *A*, **Taxiphyllum alternans,** leaves (× 21). *B–C:* **Taxiphyllum taxirameum.** *B*, leaves (× 21). *C*, leaf tip (× 328). *D–F:* **Ctenidium molluscum.** *D*, branch leaf (× 21). *E*, stem leaf (× 21). *F*, leaf tip (× 86).

1. **Taxiphyllum alternans**
(Card.) Iwats.
Figure 57, A

Plants rather robust, in soft, glossy mats; stems ± flattened, not much branched; leaves spreading, complanate, distant on older parts of stems, ovate to broadly ovate-lanceolate, acute to acuminate, to 1.6 mm wide; costa usually conspicuous, double, one branch sometimes reaching nearly ½ leaf length. Sporophyte unknown.

Habitat: Humus, rotted logs, roots, soil; swamps and wet forests.

Range: Southeastern United States; Asia.

Gulf Coast Distribution: Florida and Louisiana.

This species is easy to recognize due to its habitat and its ± robust stature. It is considerably larger in size than our other species of *Taxiphyllum* or the related genus, *Isopterygium*. It is apparently quite rare on the Gulf Coast and is known there only from a few collections in Lafayette and Vermilion parishes, Louisiana, and a few in Florida. However, it could be undercollected due to its rather obscure habitat in swamps.

Plants slender to moderate-size, in thin, spreading, glossy mats; stems prostrate, often strongly flattened, not much branched, often mostly all pointing in the same direction; leaves mostly widely spreading, complanate, usually ± widely spaced on stem, tips often ± squarrose, ovate-lanceolate to oblong-lanceolate, acuminate, to about 0.6 mm wide; costa double or lacking, one branch sometimes reaching ⅓ leaf length. Not producing sporophytes in our area.

2. Taxiphyllum taxirameum
(Mitt.) Fleisch.
Figure 57, B−C

Taxiphyllum geophilum (Aust.) Fleisch.

Habitat: Soil, rock, tree bases; in mesic forests.

Range: Eastern and southwestern United States; tropical America; Asia.

Gulf Coast Distribution: Florida; Louisiana; Texas.

This glossy little moss is usually easily recognized in the field by its mostly strongly flattened stems which often all appear to point in the same direction. The plants are generally conspicuously smaller than those of *T. alternans* and have narrower leaves.

TAXIPHYLLUM DEPLANATUM (Bruch & Schimp.) Fleisch. was attributed to Louisiana by Ireland (1969). However, the specimen upon which Ireland based his report was later annotated by him as *T. taxirameum*, which indeed it is. This species has a northerly range in North America and does not reach the Gulf Coast.

Plants medium-size, glossy, yellowish green, in dense, soft mats; stems creeping, pinnately branched, branches short; branch leaves smaller than stem leaves, slenderly lanceolate to triangular-acuminate, with filiform acumination, serrate, falcate-secund; cells linear, dorsally papillose in upper part of leaf, subquadrate in basal angles; stem leaves broadly obcordate and decurrent, abruptly long-acuminate, serrate, spreading to squarrose, ± secund, often a little plicate. Not producing sporophytes in our area.

7. CTENIDIUM (Schimp.) Mitt.

Ctenidium molluscum (Hedw.) Mitt.
Figure 57, D−F

Habitat: Banks of ravines and streams in upland forests.

Range: Europe; northern Africa; Asia; eastern North America.

Gulf Coast Distribution: Mississippi to eastern Texas.

This attractive moss is fairly common in upland forests in Louisiana and adjacent areas of Mississippi and Texas. It has not been reported from southern Alabama or the Florida Panhandle but probably occurs in both places. The squarrose, secund, slenderly acuminate leaves and the plants in soft, shiny mats are distinctive. Under the microscope the dorsally papillose upper cells and differentiated stem and branch leaves distinguish *Ctenidium* from *Hypnum*.

MITTENOTHAMNIUM DIMINUTIVUM (Hampe) E. G. Britt. has been reported from two counties in northwest Alabama but not from within our range. I have not been able to see the Alabama specimens. This moss is otherwise known in the United States from Florida and southern Georgia and could well turn up in our area, especially to the east. The plants are small, slender and prostrate, and have slenderly acuminate serrulate leaves, with the cells papillose dorsally by projecting cell ends. At first glance the plants resemble a *Sematophyllum* or *Isopterygium*; the papillose cells are distinctive.

PYLAISIELLA POLYANTHA (Hedw.) Grout and P. SELWYNII (Kindb.) Crum have both been reported from Gulf Coast sites (as *Pylaisia*), but probably do not reach this far south. I have not seen any authentic specimens of either from our area. *Pylaisiella* resembles *Platygyrium* in general but differs, among other ways, in lacking brood bodies, in having the branches curved upward, and in lacking the dark oily appearance so characteristic of *Platygyrium*. These two species characteristically grow on tree trunks, in contrast to the more typical log and stump habitat of *Platygyrium* in our area.

DIPHYSCIACEAE Fleisch.

Plants small, in dull, dark green to brownish or blackish turves; stems very short, erect, not or little-branched; lower leaves small, crispate when dry, ± erect-spreading when moist, ligulate to lingulate, obtuse to acute, sometimes cucullate at apex, margins entire to crenulate-serrulate; cells small, obscure, ± isodiametric, in 2–3 layers above, flat to strongly mammillose or bulging, basal cells unistratose, pale, rectangular; costa single, ending below or in apex; upper and perichaetial leaves much larger than lower ones, ovate-lanceolate, long and slenderly acuminate, ± ciliate and scarious above, awned by the long-excurrent costa. Setae very short; capsules immersed, strongly oblique; peristome double, exostome teeth very short and fragile, endostome a conspicuous, pale, plicate membrane; operculum slenderly conic; calyptra small, conic, naked.

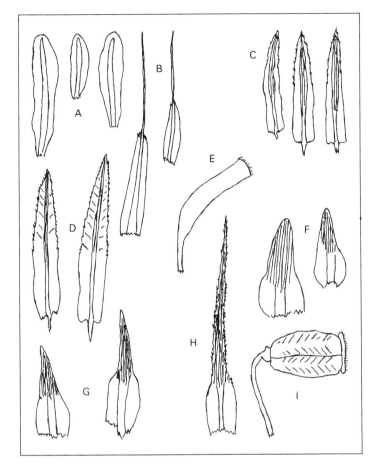

Fig. 58. *A—B*: **Diphyscium foliosum.** *A*, leaves (× 21). *B*, perichaetial leaves (× 8). *C*, **Atrichum angustatum,** leaves (× 8). *D—E*: **Atrichum undulatum.** *D*, leaves (× 5). *E*, capsule (× 5). *F*, **Pogonatum brachyphyllum,** leaves (× 8). *G*, **Pogonatum pensilvanicum,** leaves (× 8). *H—I*: **Polytrichum commune.** *H*, leaf (× 5). *I*, capsule (× 5).

The Diphysciaceae are a small family of mosses with very distinctive sporophytes. *Diphyscium foliosum* (Hedw.) Mohr has been found in our area.

Very small plants in dark green to brownish or blackish sods; fertile plants conspicuously larger than sterile ones; lower leaves (and those on sterile stems) short, lingulate, obtuse; margins crenulate-serrulate; cells obscure, mammillose; costa strong; perichaetial leaves larger than lower ones, costa excurrent as a long, brown, spinulose awn. Capsules highly oblique; operculum long-conic; exostome teeth very small, fragile, imperfect, difficult to see; endostome whitish, conspicuous. Not found with sporophytes in our area.

Habitat: Sandy banks and humus in mesic forests.
Range: Europe; Japan; eastern North America; Mexico; Guatemala; Jamaica.
Gulf Coast Distribution: Louisiana.

DIPHYSCIUM Mohr

Diphyscium foliosum
(Hedw.) Mohr
Figure 58, A—B

Diphyscium foliosum is apparently known on the Gulf Coast only from a single collection, sterile, from central Louisiana. The sods of sterile plants have a decidedly pottiaceous appearance, but the crispate, lingulate leaves and the obscure upper cells in two to three layers are distinctive. The fertile stems, if present, are easy to spot with the unaided eye and identify *D. foliosum* immediately. This species probably occurs over much of the inland area along the Gulf Coast in favorable sites. It should be sought along stream and ravine banks in shady, mesic, broadleaf forests. Sterile plants are very small and difficult to find unless growing in extensive colonies, where the dark green to brownish or blackish color is distinctive.

POLYTRICHACEAE Schwaegr.

Small to robust, firm plants in dark green to brownish or blackish, thin to dense colonies; stems erect, mostly unbranched; leaves straight and erect-appressed to crispate-contorted when dry, spreading when moist, blunt to acuminate from mostly sheathing base; margins mostly toothed, sometimes bordered with elongate cells; cells ± isodiametric, smooth or sometimes papillose; costa single, very strong, bearing several to many long, green, parallel lamellae on upper surface, the lamellae sometimes extending nearly the entire width of the upper surface of the leaf. Setae elongate; capsules cylindric or with four to six angles, erect or inclined, sometimes curved; peristome single, mostly of 32 or 64 short, firm, lingulate, multicellular teeth, the teeth attached at their tips to a flat, circular membrane (epiphragm); operculum acute to rostrate; calyptra cucullate, densely hairy or sometimes naked.

The Polytrichaceae are represented on the Gulf Coast by three genera. Members of the family are easily recognized by the lamellae on the ventral surface of the costa and by the massive peristome teeth.

1. Leaves strongly crispate-contorted when dry; lamellae few; calyptra naked — 1. ATRICHUM
1. Leaves erect to spreading-flexuous when dry, only slightly or not contorted; lamellae numerous; calyptra densely hairy — 2
  2. Plants very small, less than 1 cm tall; leaf margins mostly entire; capsule cylindric — 2. POGONATUM
  2. Plants usually robust, to 30 cm or more tall; leaf margins sharply serrate above base; capsule with 4 angles — 3. POLYTRICHUM

Medium-size to rather robust plants in dark green, dull tufts or turves; stems erect; leaves strongly crispate-contorted when dry, erect-spreading when moist, oblong to linear-lanceolate, mostly undulate above and toothed dorsally on lamina and costa, bordered by elongate cells, toothed on margins; cells ± isodiametric to wider than long, mostly smooth; costa bearing several parallel lamellae ventrally. Setae single or several per perichaetium; capsules cylindric, nearly erect to inclined, often ± curved; peristome of 32 teeth; operculum long-rostrate; calyptra mostly smooth and naked.

Two species of *Atrichum* are common along the Gulf Coast.

1. ATRICHUM P.-Beauv.

Leaves narrow; plants mostly less than 20 mm tall, often much less; lamellae conspicuous, wavy, 5–14 cells tall, covering ¼–½ width of leaf in upper ⅓ .... 1. A. ANGUSTATUM

Leaves relatively broad; plants mostly taller than 20 mm; lamellae inconspicuous, straight, 2–5 cells tall, covering less than ¼ width of leaf above .... 2. A. UNDULATUM

Stems mostly less than 20 mm tall; leaves conspicuously undulate, to 7 mm long; lamina toothed dorsally along undulations; lamellae mostly 5–6, conspicuous, wavy, lax, 5–14 cells high, covering ¼–½ width of leaf in upper ⅓; cells ± isodiametric, smooth or papillose. Capsules 2.5–6.5 mm long, erect to inclined, usually somewhat curved.

1. Atrichum angustatum
(Brid.) B.S.G.
Figure 58, C

> **Habitat:** On soil; banks in ± dry, open forests and disturbed sites.
> **Range:** Europe; Asia; eastern North America; Central America; Jamaica; Haiti.
> **Gulf Coast Distribution:** Florida to Texas.

Atrichum angustatum is usually easy to recognize by its habitat, small stature, narrow leaves, and by the conspicuous wavy lamellae covering much of the width of the leaf in the upper part. It generally grows in poorer, drier, and more open sites than our other species.

Plants often robust; stems to about 7 cm tall; leaves mostly ± undulate, especially above, to 9 mm long; lamina mostly conspicuously toothed dorsally in rows along the undulations, the teeth sometimes few and not in rows; lamellae 4–6, low, straight, 2–5 cells high, covering less than ¼ width of leaf in upper ⅓; cells mostly wider than long, smooth. Capsules 3–8 mm long, nearly erect to ± inclined, somewhat curved.

2. Atrichum undulatum
(Hedw.) P.-Beauv.
Figure 58, D–E

> **Habitat:** On soil in mesic to moist forests; ravines and stream banks.
> **Range:** Widely distributed in the Northern Hemisphere.
> **Gulf Coast Distribution:** Florida to eastern Texas.

Plants of this species are often tall and handsome, especially when seen moist in the field. The leaves usually have definite undulations in the upper part with conspicuous teeth (as seen under the microscope) along the undulations dorsally. *Atrichum cylindricum* (Willd. ex Weber) G. Sm. was described long ago from Louisiana and recently recognized as a distinct species. However, it is difficult to separate from *A. undulatum*, even when fertile, and Crum and Anderson (1981) include it in the concept of *A. undulatum* under the name *A. undulatum* var. *attenuatum* B.S.G. It differs from typical *A. undulatum* in that the leaves lack conspicuous undulations and the dorsal teeth are lacking or few and not in rows along the undulations.

2. POGONATUM P.-Beauv.

Plants small (our species) to robust, dark green to brownish or reddish, in thin to dense colonies, with persistent, green protonemata usually conspicuous; stems erect; leaves erect and not or little contorted when dry, with blunt to acuminate upper part from sheathing base; lamellae numerous, covering most of width of leaf above base, margins entire to toothed. Commonly producing sporophytes in our area; capsules cylindric, erect to inclined; peristome of 32 teeth; operculum conic to rostrate; calyptra densely hairy.

Two species of *Pogonatum* occur along the Gulf Coast.

Leaves broad, apices bluntly rounded	1. P. BRACHYPHYLLUM
Leaves slender, apices acuminate	2. P. PENSILVANICUM

1. **Pogonatum brachyphyllum**
(Michx.) P.-Beauv.
Figure 58, F

Plants small, scattered in persistent green protonemata; stems very short, erect; leaves broad and blunt, 1–3 mm long, erect-appressed when dry, not contorted or flexuous, margins entire above base. Calyptra densely covered with brownish hair.

> **Habitat:** Soil on bare banks; mesic to dry forests; disturbed sites.
> **Range:** Eastern United States.
> **Gulf Coast Distribution:** Florida to eastern Texas.

This species is rather common over much of our area. Its habitat is similar to that of *Atrichum angustatum*, with which it is often found growing. Its short, broad leaves and brown calyptra distinguish it easily from *P. pensilvanicum*.

2. **Pogonatum pensilvanicum**
(Hedw.) P.-Beauv.
Figure 58, G

Plants small, scattered in persistent green protonemata; stems erect, short; leaves slender, acuminate, 2–4 mm long, tips ± flexuous when dry, margins entire to irregular or ± serrate above base. Calyptra densely covered with whitish or pale yellow hair.

> **Habitat:** Soil on bare banks; ravines and along streams; mesic forests.

Range: Eastern North America; Mexico; South America.
Gulf Coast Distribution: Florida to eastern Texas.

Pogonatum pensilvanicum is less frequently encountered than *P. brachyphyllum* in our area and occurs in more mesic sites. See comments under *P. brachyphyllum*.

Plants robust, rigid, dark green to brownish, in loose to dense tufts or turves; stems erect, to 30 cm or more tall; leaves erect-appressed to spreading when dry, spreading when moist, to 12 mm long, acuminate from a pale, shiny, sheathing base; margins serrate above base; costa excurrent; lamellae numerous, covering essentially all the leaf above the base. Rarely producing sporophytes in our area; capsules with four angles, to 5 mm long; peristome of 64 teeth; operculum rostrate; calyptra densely hairy.

> **Habitat:** Acid, wet, usually sandy sites; roadsides, wet pine forests.
> **Range:** Nearly cosmopolitan.
> **Gulf Coast Distribution:** Florida to eastern Texas.

Polytrichum commune, sometimes called hair-cap moss because of its calyptra, is not common in our area for the most part; it is absent from the coastal prairies of Louisiana and Texas. Its Gulf Coast distribution coincides in general with that of pine and *Sphagnum*. The habitat and usually robust stature make this moss easy to recognize. In *Atrichum* the lamellae are much fewer, and the leaves are crispate when dry; in *Pogonatum* the stems are very short and the capsules are cylindric. Both *Atrichum* and *Pogonatum* commonly produce sporophytes in our area, in contrast to *Polytrichum*.

3. POLYTRICHUM Hedw.

Polytrichum commune Hedw.
Figure 58, H–I

Glossary

ABAXIAL. Away from the axis; *see also* DORSAL.

ACROCARPOUS. Producing the archegonia and sporophytes at the tips of the stems. *Acrocarpous mosses* are mostly erect plants growing in tufts or sods, rather than prostrate mat-formers.

ACUMEN. A narrow, tapering point.

ACUMINATE. Gradually drawn out to a slender tip.

ACUTE. Sharply pointed—less than a 90° angle.

ADAXIAL. Toward the axis; *see also* VENTRAL.

ALAR. In mosses, referring to the basal angles of a leaf. The *alar cells* are often differentiated from those of the rest of the leaf.

ANNULUS. The ring of differentiated cells connecting the operculum to the urn; aiding in dehiscence and dehiscent itself in some mosses.

ANTHERIDIUM. The male reproductive organ; produces the sperm.

APICAL. At the tip.

APICULATE. Abruptly short-pointed.

APICULUS. A short, sharp point.

APPRESSED. Closely applied to one another or to the stem; mostly referring to leaves.

ARCHEGONIUM. The female reproductive organ; produces the egg.

AREOLATION. The network of cells in a leaf.

ATTENUATE. Slenderly drawn out.

AURICLE. An earlike lobe on the basal margin of a leaf.

AURICULATE. With auricles.

AUTOICOUS. With the archegonia and antheridia in separate inflorescences on the same plant.

AWN. A slender, hairlike or bristlelike point, usually formed by the excurrent costa.

AXIL. The angle between a leaf and the stem.

AXILLARY. In a leaf axil.

BEAK. The elongate apex of an operculum or calyptra.

BIPLICATE. Having two longitudinal folds.

BISTRATOSE. In two layers.

BORDERED. With the margins differentiated; the marginal cells of a different color, texture, shape, or thickness than those of the rest of the leaf.

BRACTS. Modified leaves around the groups of sex organs.

BROOD BODIES. Specialized structures serving in asexual reproduction, including gemmae, propagula, and tubers.

BULBIFORM. Shaped like a bulb.

BULBIL. A small, dehiscent bud serving in asexual reproduction; a type of propagulum.

CALCAREOUS. Containing calcium carbonate, as limestone.

CALCICOLOUS. Living in calcareous habitats.

CALCIPHILE. A plant characteristically occurring on calcareous substrates such as limestone, concrete and mortar.

CALYPTRA. The usually delicate and quickly falling structure covering at least the upper part of the capsule; derived from the archegonium, which enlarges and then is torn away and carried up on top of the capsule as the seta elongates.

CAMPANULATE. Bell-shaped.

CANCELLINAE. The conspicuous areas of large, empty, hyaline, porose cells in the leaf bases of mosses of the family Calymperaceae.

CAPITATE. With a headlike terminal cluster of leaves or leaves and short branches.

CAPSULE. The spore-bearing portion of the sporophyte.

CHANNELED. Trough-shaped.

CILIA. Slender, hairlike features alternating with the segments of the endostome in many mosses; also applied to hairlike features on leaf margins and the bases of calyptrae.

CILIATE. Bearing cilia.

CIRCINATE. Curved in a circle; coiled with the apex at the center.

CLEISTOCARPOUS. Lacking an operculum, the capsule thus rupturing irregularly to release the spores. *Cleistocarpous mosses* generally have very short setae and immersed capsules.

COLLENCHYMATOUS. Referring to cells with their walls thickened at the corners.

COLUMELLA. The central column of sterile tissue in a capsule.

COMA. A collection of larger leaves concentrated at the stem tip.

COMPLANATE. Flattened.

CONCOLOROUS. Of the same color.

CORTEX. The outer layers of cells of a stem when differentiated from the inner central cylinder or strand.

CORTICAL. Referring to the cortex.

CORTICOLOUS. Growing on bark.

COSTA. The midrib of a moss leaf; sometimes called the nerve or vein.

COSTATE. With a costa.

CRENATE. With rounded projections or teeth.

CRENULATE. With small, rounded teeth.

CRISPATE. Wavy; highly contorted.

CUCULLATE. Hood-shaped, as a calyptra split along one side; also applied to leaves very concave at the tips.

CUSPIDATE. With an abrupt point; longer than apiculate.

DECIDUOUS. Regularly falling away.

DECURRENT. With the margins or costa, or both, extending down the stem below the leaf insertion as wings or ridges, these usually tearing off with the leaf.

DEHISCENT. Splitting open to release the spores, by valves, operculum, or irregularly.

DELTOID. Equilaterally triangular.

DENDROID. Resembling a tree in form; *dendroid mosses* look like miniature trees, *e.g., Climacium*.

DENTATE. With outward-pointing teeth.

DENTICULATE. Finely dentate.

DIMORPHIC. With two forms.

DIOICOUS. With antheridia and archegonia borne on separate plants; having male plants and female plants.

DISTICHOUS. In two rows, the rows opposite one another on the stem.

DORSAL. The abaxial surface of a leaf; the upper surface of a prostrate plant.

ECHINATE. With spines.

ECOSTATE. Lacking a costa.

EMARGINATE. Shallowly notched at the apex.

EMERGENT. Partly exposed; said of capsules that are only partly concealed by the perichaetial leaves.

ENDOSTOME. The inner series of a double peristome; composed of the segments, which often alternate with cilia; sometimes reduced to a membrane; often arising from a basal membrane.

ENTIRE. Lacking projections.

EPIPHRAGM. The membrane connecting the tips of the peristome teeth in the family Polytrichaceae.

EPIPHYTE. A plant that grows upon another plant.

EPIPHYTIC. Growing on a plant.

EXCURRENT. Projecting beyond the apex of the lamina.

EXOSTOME. The outer series of a double peristome; composed of the teeth.

EXOTHECIAL. Referring to the *exothecium*.

EXOTHECIUM. The outermost layer of the capsule wall; composed of the *exothecial cells*.

EXSERTED. Exposed; said of capsules that extend beyond the tips of the perichaetial leaves.

FALCATE. Sickle-shaped; curved like the blade of a sickle.

FILIFORM. Threadlike.

FIMBRIATE. Fringed.

FLAGELLIFORM. Long and slender; usually referring to slender, differentiated stems and branches bearing reduced leaves.

FLEXUOUS. Wavy.

FOLIOSE. Leaflike.

FOOT. The lowermost portion of the sporophyte; embedded in the gametophore.

FRONDOSE. Resembling a fern leaf.

FUGACIOUS. Falling away early or quickly; not persisting.

FUSIFORM. Tapering to both ends; spindle-shaped.

GAMETE. A sex cell; sperm and egg.

GAMETOPHORE. The leafy, gamete-bearing plant; part of the gametophyte generation of the moss life cycle.

GAMETOPHYTE. The gamete-producing phase in the moss life cycle; includes the protonema and the gametophore.

GEMMAE. (*sing.* GEMMA) Small, few-celled asexual reproductive structures; mostly ± spherical or filamentous.

GEMMIFEROUS. Bearing gemmae.

GLAUCOUS. With a whitish, bluish, or grayish coating or appearance, as the waxy bloom of grapes and plums.

GREGARIOUS. Growing close together but not forming turves or cushions.

GUIDE CELLS. Large, empty cells in a median row in the costa, as seen in cross section; usually accompanied by stereid cells.

HAIR-POINT. A slender, hairlike leaf tip, usually formed by a long-excurrent costa.

HOMOMALLOUS. Pointing in one direction.

HYALINE. Colorless and ± transparent.

HYGROSCOPIC. Responding to changes in moisture by changing shape or position, or both.

HYPOPHYSIS. An enlargement of the neck of a capsule.

IMBRICATE. Overlapping like shingles, and appressed.

IMMERSED. Covered; concealed; the capsule exceeded by the tips of the perichaetial leaves; also said of sunken stomata.

INCRASSATE. Thickened.

INFLATED. Swollen; enlarged.

INFLORESCENCE. A cluster of sex organs with the surrounding leaves or bracts.

INNOVATION. A new shoot or branch, especially one arising at the base of an inflorescence.

INOPERCULATE. Lacking a defined operculum.

INTERNODE. The interval of stem between adjacent nodes.

INVOLUTE. Inrolled.

ISODIAMETRIC. About as long as broad.

ISOPHYLLOUS. Stem and branch leaves similar (*Sphagnum*).

JULACEUS. Smooth and cylindric; said of stems and branches with imbricate leaves.

KEELED. Sharply folded.

LACERATE. Irregularly deeply torn.

LAMELLAE. (*sing.* LAMELLA) Thin plates or ridges arising from the costa in some mosses.

LAMINA. The leaf blade, as distinct from the costa.

LANCEOLATE. Like a lance head; broadest below midleaf and tapering to a point above.

LAX. Loose; usually referring to the nature and arrangement of cells in the leaf; large, thin-walled cells.

LIGULATE. Strap-shaped; narrower than lingulate.

LINEAR. Long and narrow with parallel sides; narrower than ligulate.

LINGULATE. Tongue-shaped; broader than ligulate.

LOESS. Soft, very fine-grained soil deposited by wind.

LUMEN. The cell cavity.

MAMMILLOSE. Bulging and with a nipplelike projection.

MEDIAN. Middle; median cells are those about $1/2 - 2/3$ up the leaf, between the costa (if present) and the margin.

MICROPHYLLOUS. With small leaves.

MITRATE. Conic; entire at base, or with two or more lobes; not deeply split along one side.

MONOICOUS. With antheridia and archegonia borne on the same plant.

MUCRO. An abrupt, short point.

MUCRONATE. With a mucro.

MULTIFORM. Of more than one form.

NECK. The often differentiated basal portion of a capsule.

OBCONICAL. Cone-shaped, with the pointed end of the cone downward.

OBCORDATE. Heart-shaped, with the broad end downward.

OBOVATE. Egg-shaped in outline, with the broad end uppermost.

OBTUSE. Blunt; broadly pointed, the tip forming an angle of more than 90°.

OPERCULATE. With an operculum.

OPERCULUM. The lid of the capsule; its dehiscence allows the spores to escape.

ORBICULAR. Approximately circular in outline.

OVATE. Egg-shaped in outline, with the broad end downward.

OVOID. Egg-shaped.

PAPILLA. (*pl.* PAPILLAE) A small protuberance.

PAPILLOSE. With papillae.

PARAPHYLLIA. Small structures of various forms arising from the stem among the leaves; not associated with branch buds or bases. *See* PSEUDOPARAPHYLLIA.

PARAPHYSES. Multicellular, hairlike filaments produced among the antheridia and sometimes among the archegonia.

PARENCHYMATOUS. Composed of ± short cells joined end-to-end, the ends usually broad and not overlapping appreciably. *See* PROSENCHYMATOUS.

PAROICOUS. With antheridia and archegonia in the same inflorescence but not mingled; the antheridia axillary in the upper and perichaetial leaves, below the archegonia.

PELLUCID. Clear.

PENDENT. Hanging.

PERCURRENT. Extending to the apex.

PERICHAETIAL. Referring to the perichaetium.

PERICHAETIUM. The archegonial inflorescence.

PERIGONIAL. Referring to the perigonium.

PERIGONIUM. The antheridial inflorescence.

PERISTOME. The one or two series of teeth around the mouth of the capsule; the members of the outer row are *teeth*; those of the inner row are *segments*, sometimes alternating with cilia.

PILOSE. With long hairs.

PINNATE. With numerous, usually close, spreading, ± equal branches on opposite sides of the stem; in outline, the stem with its branches often resembling a feather.

PIT. A small depression or pore in a cell wall.

PLICATE. Folded in longitudinal pleats, like a fan.

PLURIPAPILLOSE. With more than one papilla.

POLYMORPHOUS. With more than one form; variable.

PORE. An opening in a cell wall.

POROSE. With pores; pitted.

PRIMARY STEM. The main stem, from which the branches or secondary stems arise; usually prostrate and creeping; sometimes inconspicuous.

PROPAGULA. (*sing.* PROPAGULUM) Asexual reproductive bodies; reduced buds, branchlets, or leaves; deciduous.

PROPAGULIFEROUS. Bearing propagula.

PRORULOSE. With the tips of the cells protruding over the ends of the cells above or below, or both; the cell tips thus resembling the prow of a ship.

PROSENCHYMATOUS. With narrow, elongate cells whose ends overlap one another. *See* PARENCHYMATOUS.

PROTONEMA. (*pl.* PROTONEMATA) The filamentous, algalike phase of the moss gametophyte that develops from the spores, later giving rise to the leafy gametophores.

PSEUDOPARAPHYLLIA. Very small filamentous or leaflike appendages of the stem associated with branch buds or bases in pleurocarpous mosses. *See* PARAPHYLLIA.

PSEUDOPODIUM. An elongated, usually naked extension of the stem, functioning as a seta in *Sphagnum* and *Andreaea*; sometimes applied to similar features bearing brood bodies in various mosses.

PYRIFORM. Pear-shaped.

QUADRATE. Square.

RADICLES. *See* RHIZOIDS. Sometimes applied more to conspicuously matted rhizoids on the lower portions of a plant.

RADICULOSE. Bearing radicles.

RECURVED. Curved downward or backward.

REFLEXED. Bent backward.

RESORPTION FURROW. A groove along the leaf margins in some species of *Sphagnum*, formed by the dissolution of the outer walls of the marginal cells.

RETICULATE. Netted.

REVOLUBLE. Applied to an annulus which falls away at maturity.

REVOLUTE. Rolled downward.

RHIZOIDS. Slender, multicellular, threadlike filaments attaching the moss plant to the substrate; sometimes matted on the lower portions of the plant.

RHIZOME. An elongate, slender, prostrate, subterranean stem; the primary stem of many mosses is rhizomelike.

RHOMBIC. Diamond-shaped.

RHOMBOIDAL. Longer than rhombic.

ROSTRATE. With a slender beak, the *rostrum*.

ROSTRUM. A slender beak.

RUGOSE. Transversely wrinkled.

SCARIOUS. Dry and membranous.

SECONDARY STEM. A stemlike branch arising from a primary stem and often itself branched.

SECUND. Strongly homomallous; conspicuously pointing in the same direction.

SEGMENTS. The major divisions of the endostome; often alternating with more slender divisions, the cilia.

SEPTATE. With partitions.

SERIALLY. In rows.

SERPENTINE. Curved like a snake.

SERRATE. Toothed, with the teeth pointing forward.

SERRULATE. Finely toothed.

SESSILE. Lacking a stalk or seta.

SETA. The usually slender and elongate stalk that bears the capsule; part of the sporophyte.

SETACEOUS. Bristlelike.

SILICEOUS. With a high content of silicon; sandy, or sandy in origin, as sandstone.

SIMPLE. Not branched.

SINUOSE. Wavy.

SORDID. Having a dull, more or less muddy color.

SPATULATE. Spatula-shaped; broad above, abruptly narrowed downward.

SPINOSE. Spiny.

SPOROPHYTE. The spore-producing phase in the moss life cycle; consisting of the foot, seta, and capsule.

SQUARROSE. Spreading at right angles.

STEGOCARPOUS. With a differentiated operculum.

STEREIDS. Thick-walled cells with small lumens (as seen in cross section); in the costae of many mosses.

STIPE. The unbranched, stalklike lower portion of the (usually) secondary stem of some dendroid or frondose mosses.

STOLONIFORM. Referring to slender, elongated, creeping stems and branches, usually bearing reduced leaves and with radicles at the tips.

STRIATE. With fine, longitudinal ridges.

STRICT. Straight and rigid.

SUBPINNATE. Nearly pinnate; approximating the pinnate condition.

SUBULA. A long, slender point.

SUBULATE. With a subula; narrowly long-acuminate.

Sᴜʟᴄᴀᴛᴇ. Strongly folded longitudinally; grooved.

Sʏɴᴏɪᴄᴏᴜs. With antheridia and archegonia mixed in the same inflorescence.

Tᴀxᴏɴ. A taxonomic unit; family, genus, species, variety, etc.

Tᴇᴇᴛʜ. Divisions of the exostome in mosses with a double peristome, or the divisions of a single peristome.

Tᴇʀᴇᴛᴇ. Circular in cross section.

Tᴇʀᴍɪɴᴀʟ. At the end or tip.

Tᴏᴍᴇɴᴛᴏsᴇ. Woolly.

Tᴏᴍᴇɴᴛᴜᴍ. Matted rhizoids.

Tᴏᴏᴛʜ. One of the divisions of the exostome, or of a single peristome; also referring to projections from the margins of leaves.

Tʀᴜɴᴄᴀᴛᴇ. Appearing to be cut off squarely at the tip.

Tᴜʙᴇʀ. The small, ± globose, multicellular, subterranean bodies produced on the rhizoids of many kinds of mosses; often colored; sometimes called *rhizoid gemmae*.

Tᴜʙᴇʀᴄᴜʟᴀᴛᴇ. Roughened with prominent projections.

Tᴜʙᴜʟᴏsᴇ. Tubelike, as leaves with strongly incurved margins.

Tᴜꜰᴀ. A porous form of limestone, often formed by deposition of calcareous materials around mosses growing in water.

Tᴜᴍɪᴅ. Swollen.

Uᴍʙᴏɴᴀᴛᴇ. Convex, with a blunt central projection.

Uɴᴅᴜʟᴀᴛᴇ. Wavy.

Uɴɪᴘᴀᴘɪʟʟᴏsᴇ. With a single papilla.

Uɴɪsᴇʀɪᴀᴛᴇ. In a single row.

Uɴɪsᴛʀᴀᴛᴏsᴇ. In a single layer.

Uʀɴ. The spore-bearing portion of the capsule.

Vᴀɢɪɴᴀᴛᴇ. Sheathing; in *Fissidens* the vaginant laminae of the leaf clasp the stem.

Vᴇɴᴛʀᴀʟ. Referring to the lower surface of a prostrate plant or organ and to the adaxial (upper) surface of a leaf.

Vᴇʀᴍɪᴄᴜʟᴀʀ. Worm-shaped.

Vᴇʀʀᴜᴄᴏsᴇ. Warty.

Literature
Cited

Anderson, L. E. 1954. Hoyer's solution as a rapid permanent mounting medium for bryophytes. *Bryologist* 57:242–44.

Anderson, L. E., and Bryan, V. S. 1958. Systematics of the autoicous species of *Ditrichum* subg. *Ditrichum*. *Brittonia* 10:121–37.

Andrus, R. E. 1974. Significant new distributional records for the genus *Sphagnum* in the northeastern United States. *Rhodora* 76:511–18.

———. 1979. *Sphagnum bartlettianum* in the southeastern United States. *Bryologist* 82:198–203.

Bent, A. C. 1948. Life histories of North American nuthatches, wrens, thrashers and their allies. Order Passeriformes. Smithsonian Institution. Bulletin of the United States National Museum 195:i–xi, 1–475.

Black, J. M., ed. 1943–57. *Flora of south Australia.* 2nd ed. Pp. 342, 531. Adelaide, Australia.

Braithwaite, R. 1880. *The Sphagnaceae or peat-mosses of Europe and North America.* London.

Breen, R. S. 1963. *Mosses of Florida: An illustrated manual.* Gainesville.

Breil, D. A. 1970. Liverworts of the Mid-Gulf coastal plain. *Bryologist* 73:409–91.

Breil, D. A., and Moyle, S. M. 1976. Bryophytes used in construction of bird nests. *Bryologist* 79:95–98.

Brotherus, V. F. 1924–25. Musci. In *Die Natürlichen Pflanzenfamilien*, by A. Engler and K. Prantl, vol. 10–11. 30 vols. to date. Leipzig.

Conard, H. S. 1979. *How to know the mosses and liverworts.* Rev. by P. L. Redfearn. Dubuque, Iowa.

Crum, H. 1972. The geographic origins of the mosses of North America's eastern deciduous forest. *Journal of the Hattori Botanical Laboratory* 35:269–98.

———. 1976. *Mosses of the Great Lakes Forest.* Rev. ed. Ann Arbor.

Crum, H., and Anderson, L. E. 1981. *Mosses of eastern North America.* New York.

Crundwell, A. C., and Nyholm, E. 1964. The European species of the *Bryum microerythrocarpum* complex. *Transactions of the British Bryological Society* 4(4):597–637.

Delgadillo, M. C. 1975. Taxonomic revision of *Aloina, Aloinella* and *Crossidium* (Musci). *Bryologist* 78:245–303.

Fife, A. J. 1979. Taxonomic observations on three species of North American Funariaceae. *Bryologist* 82:204–14.

Frahm, J.-P. 1981. Ein praktisches Einschlussmittel für Mikropräparate von Moosen. *Herzogia* 5: 531–33.

Frye, T. C., and Clark, L. 1937–47. *Hepaticae of North America*. Seattle.

Grout, A. J. 1928–45. *Moss flora of North America north of Mexico*. Newfane, Vt.

Hedwig, J. 1801. *Species Muscorum Frondosorum*. Leipzig.

Ireland, R. R. 1969. A taxonomic revision of the genus *Plagiothecium* for North America, north of Mexico. National Museums of Canada, National Museum of Natural Sciences, Publications in Botany 1:i–viii, 1–118.

Linnaeus, C. 1753. *Species Plantarum*. Stockholm.

Magill, R. E. 1973. Notes on Texas mosses II. *Bryologist* 76:557–59.

Ochi, H. 1974. Some bryaceous "Old World" mosses, also distributed in the New World. *Journal of the Faculty of Education, Tottori University: Natural Science* 25(1–2):35–41.

Redfearn, P. L., Jr. 1972. Mosses of the interior highlands of North America. *Annals of the Missouri Botanical Garden* 59:1–104.

Rogers, K. E., and Griffin, D. G., III. 1974. Notes on Mississippi bryophytes I. *Castanea* 39:239–62.

Saito, K. 1975. A monograph of Japanese Pottiaceae (Musci). *Journal of the Hattori Botanical Laboratory* 39:373–537.

Sayre, G. 1971. Cryptogamae exsiccatae: An annotated bibliography of published exsiccatae of Algae, Lichens, Hepaticae, and Musci. IV. Bryophyta. *Memoirs of the New York Botanical Garden* 19(2):175–276.

Schuster, R. M. 1966–. *Hepaticae and Anthocerotae of North America east of the hundredth meridian*. 4 vols. to date. New York.

Smith, A. J. E. 1978. *The moss flora of Britain and Ireland*. Cambridge.

Smith, A. J. E., and Whitehouse, H. L. K. 1978. An account of the British species of the *Bryum bicolor* complex including *B. dunense* sp. nov. *Journal of Bryology* 10(1):29–47.

Snider, J. A. 1975. A revision of the genus *Archidium* (Musci). *Journal of the Hattori Botanical Laboratory* 39:105–201.

Stoneburner, A., and Wyatt, R. 1979. Three Big Thicket bryophytes new to Texas. *Bryologist* 82: 491–93.

Syed, H. 1973. A taxonomic study of *Bryum capillare* Hedw. and related species. *Journal of Bryology* 7(3):265–326.

Taylor, A. M. 1927. Some ecological habitats in the longleaf pine flats of Louisiana. *Bulletin of the Torrey Botanical Club* 54:155–72.

Thieret, J. W. 1965. Bryophytes as economic plants. *Economic Botany* 10:75–91.

Vitt, D. H. 1973. A revision of the genus *Orthotrichum* in North America, north of Mexico. *Bryophytorum Bibliotheca* 1:1–208.

Welch, W. H. 1960. *A monograph of the Fontinalaceae*. The Hague.

Whitehouse, E., and McAllister, F. 1954. The mosses of Texas. A catalogue with annotations. *Bryologist* 57:63–146.

Whitehouse, H. L. K. 1966. The occurrence of tubers in European mosses. *Transactions of the British Bryological Society* 5(1):103–16.

———. 1973. The occurrence of tubers in *Pohlia pulchella* (Hedw.) Lindb. and *Pohlia lutescens* (Limpr.) Lindb. fil. *Journal of Bryology* 7(4):533–40.

Wilkes, J. C. 1963. Some mosses from Mississippi. *Bryologist* 66:204–208.

———. 1965. A check list of the mosses of Alabama. *Journal of the Alabama Academy of Sciences* 36:75–93.

Zales, W. M. 1973. A taxonomic revision of the genus *Philonotis* for North America, north of Mexico. Ph.D. dissertation, University of British Columbia, Vancouver.

Zander, R. H. 1979. Notes on *Barbula* and *Pseudocrossidium* (*Bryopsida*) in North America and an annotated key to the taxa. *Phytologia* 44(4):177–214.

Taxonomic Index

Names of taxa formally treated in this book are in boldface, and the page numbers for the main entry for these taxa are also in boldface. Synonyms and other names are in roman type.

Rhizogoniaceae 148
Rhizogonium spiniforme 148
Rhynchostegium serrulatum 213

Schizomitrium pallidum 181
Schlotheimia 158, 162
   rugifolia 137, 161, 162
Schwetschkeopsis 183, 186
   denticulata 186
   fabronia 186
Sciaromium lescurii 206
Sematophyllaceae 217
Sematophyllum 218, 227, 230
   adnatum 93, 187, 218, 220,
   adnatum 93, 187, 218, 220, 227
   caespitosum 220
   carolinianum 219
   demissum 219, 220
Silver moss 140
Solmsiella kurzii 154
Sphagnaceae 34
Sphagnum 35, 169, 235
   bartlettianum 46
   compactum 40, 46
   cuspidatum 50, 51
   cuspidatum var. serrulatum 51
   cuspidatum var. torreyanum 51
   cyclophyllum 38, 45
   erythrocalyx 41
   fitzgeraldii 49
   henryense 43
   imbricatum 42, 43
   lescurii 38, 43, 44, 49
   macrophyllum 38
   macrophyllum var. floridanum 39
   magellanicum 40, 46
   molle 47, 51

palustre 35, 43
perichaetiale 40, 41
plumulosum 51
portoricense 41
recurvum 48
strictum 45, 46
subnitens 51
subsecundum 45
tabulare 47
tenerum 48
torreyanum 51
trinitense 51
warnstorfii 47
Splachnaceae 118, 134
Splachnobryum 97, 118
   bernoulii 118
   obtusum 118
Splachnum pennsylvanicum 134
Stereophyllum radiculosum 217
Syrrhopodon 92
   floridanus 94
   incompletus 93
   incompletus var. incompletus 94
   ligulatus 95
   parasiticus 92, 95
   prolifer 94, 179
   prolifer var. papillosus 95
   texanus 92, 94, 179

Taxiphyllum 227
   alternans 217, 228, 229
   deplanatum 229
   geophilum 229
   mariannae 227
   taxirameum 228, 229
Tetraplodon pennsylvanicum 134
Thelia 188

asprella 189, 190
hirtella 189, 190
lescurii 189, 190
Thuidiaceae 188, 197
Thuidium 196, 198, 199
   allenii 201, 202, 203
   delicatulum 189, 200, 201, 202, 203
   minutulum 198, 200, 201
   pygmaeum 201
   recognitum 203
Tortella 97, 102, 106
   flavovirens 107
   humilis 107, 108
Tortula 97, 115
   muralis 117
   pagorum 116, 117
   rhizophylla 112, 117
Trematodon longicollis 81
Trichostomum 97, 106
   cylindricum 106
   jamaicense 102
   molariforme 106
   tenuirostre 106
Tuerckheimia angustifolia 97, 105

Weissia 97, 99, 101, 158
   andrewsii 103
   controversa 99, 101, 102, 103, 112
   controversa var. longiseta 103
   hedwigii 103
   jamaicense 102, 103, 107
   microstoma 103